ADENOVIRUSES

Edited by **Yulia Desheva**

Adenoviruses
http://dx.doi.org/10.5772/intechopen.74757
Edited by Yulia Desheva

Contributors

Babita Agrawal, Shakti Singh, Rakesh Kumar, Angel S Galabov, Verónica Martín, José M Rojas, Noemí Sevilla, Yulia A. Desheva

First published in London, United Kingdom, 2019 by IntechOpen
IntechOpen is the global imprint of INTECHOPEN LIMITED, registered in England and Wales, registration number: 11086078, The Shard, 25th floor, 32 London Bridge Street
London, SE19SG – United Kingdom
Printed in Croatia

British Library Cataloguing-in-Publication Data
A catalogue record for this book is available from the British Library

Additional hard copies can be obtained from orders@intechopen.com

Adenoviruses, Edited by Yulia Desheva
p. cm.
Print ISBN 978-1-78984-990-5
Online ISBN 978-1-78984-991-2

Meet the editor

Dr. Yulia Desheva is a leading researcher at the Institute of Experimental Medicine, St. Petersburg, Russia, and she holds the position of professor at St. Petersburg State University. She has her expertise in the preparation of vaccine strains for live influenza vaccines, the development of immunization schemes for high-risk individuals, and the evaluation of neuraminidase antibodies in influenza infection and vaccination. The study of molecular mechanisms of viral–bacterial interactions in the development of associated vaccines against bacterial complications of influenza infection is also in her area of interest.

Contents

Preface

Adenovirus infection is a disease of the group of acute respiratory viral infections and is characterized by lesions of the mucous membranes of the upper respiratory tract, conjunctiva, and lymphoid tissue. Adenovirus infection is characterized by widespread prevalence: up to 10% of all human viral diseases account for these viruses. The greatest susceptibility to adenoviruses is demonstrated by children from 6 months to 5 years. In some cases, there is a long persistence of adenoviruses in the human body and the transition to a chronic form of infection. A number of serotypes of adenoviruses induce tumors in animals. Adenovirus infection is especially dangerous for persons with immunodeficiency, including AIDS patients and recipients of bone marrow and internal organs, in whom adenoviruses cause severe generalized infection. At the same time, adenoviruses are considered to be one of the most studied models of oncolytic viruses and viral vectors, since these DNA-containing viruses are convenient for genetic engineering manipulations, are relatively low pathogenically, and can grow well in cell culture.

This book includes a series of articles that highlight a number of issues of human and animal adenoviruses, progress in approaches to the chemotherapy of adenoviral infections, as well as advances in the development of vector vaccines and adenovirus-based gene therapy.

I hope that this book will stimulate additional interest in this field of research and will be useful not only for researchers, students, and clinical interns, but also for practicing healthcare system doctors.

I would like to thank all the excellent authors who participated in the writing of this collection of chapters, as well as my students and members of my family who provided me with invaluable support while preparing this book. I also express appreciation to the prominent team of publishers for their clear and well-coordinated activities and assistance at all stages of this work.

Prof. Dr. Yulia Desheva
Federal State Budgetary Scientific Institution "Institute of Experimental Medicine"
Saint Petersburg, Russian Federation

Saint Petersburg State University
Saint Petersburg, Russian Federation

Chapter 1

Introductory Chapter: Human Adenoviruses

Yulia Desheva

Additional information is available at the end of the chapter

http://dx.doi.org/10.5772/intechopen.82554

1. Introduction

The date of adenoviruses' discovery is considered to be in 1953, when a cytopathogenic agent was identified during the long-term cultivation of the tissues of the tonsils and adenoids after operations in children with Rowe and coworkers [1]. This determined the name of the viruses (adenoid degeneration viruses) and outlined their basic ecology associated with asymptomatic persistence in the lymphoid tissue. Soon, adenoviruses were isolated from materials obtained from patients with acute respiratory diseases accompanied by conjunctivitis [2]. In 1954, Huebner received the new data indicating that similar viruses are also found in the secretes of patients with acute pharyngitis and conjunctivitis, and therefore they were called "adenoid-pharyngeal-conjunctival viruses" [1]. In the same year, another group of researchers, when studying the etiology of acute respiratory infections and atypical pneumonia, isolated a previously unknown virus from the US Army recruits, named RI-67. It further proved the adenoviruses' identity with the adenoid-pharyngeal-conjunctival virus [3]. In subsequent years, such viruses were isolated from patients during outbreaks of epidemic keratoconjunctivitis, although as an independent disease, it was described in the 20s of the twentieth century.

Adenoviruses are the first respiratory viruses that were isolated on tissue culture. The opportunity to grow up in vitro on synthetic media of various cells of organs and tissues of humans and animals, as well as the ability of viruses to multiply on sensitive cells to cause a cytopathic effect opened up broad prospects for the development of virology. By 1956, a large number of biologically similar but antigenically distinct strains of viruses were identified, which were decided to have a group name of adenoviruses [4]. Not less than 120 viruses that infect mammals, birds, reptiles, amphibians, and fish are described in the *Adenoviridae* family. Further study of adenoviruses, the discovery of their new serotypes made it possible to establish that this viruses cause not only respiratory diseases but also diarrhea, mesadenitis, hemorrhagic cystitis, and other pathological conditions. Only in humans, over 50 adenoviral serotypes are known which cause a wide range of illnesses, from mild respiratory infections in young children to life-threatening multi-organ disease in immunocompromised people.

2. Adenoviruses' nomenclature

This family includes two genera designated based on genetic criteria (*Atadenovirus* and *Siadenovirus*) as well as three genera, in which adenoviruses are combined according to the type of host (*Aviadenovirus*, *Ichtadenovirus*, and *Mastadenovirus*) (https://talk.ictvonline.org/taxonomy).

Genus *Atadenovirus* presents the newly formed genus which combines adenoviruses with a high relative content of AT-pairs in genomic DNA. This genus includes also adenoviruses of snakes, possums, calves, chameleon, ducks, and lizard.

Genus *Siadenovirus* combines adenoviruses containing at the 5′ end which contains a gene of an enzyme sialidase that cuts off sialic acid residues from the surface glycoproteins of the host cells. These adenoviruses infect frogs and birds.

Genus *Aviadenovirus* includes adenoviruses of turkeys, quail, chickens, and a number of other birds. The type species causes the death of embryos and respiratory disease in quails and chickens. Other members of this genus are pathogens of the egg drop syndrome, hemorrhagic enteritis, and hepatitis.

Genus *Ichtadenovirus* has the only characterized representative which infects white sturgeon—*Sturgeon ichtadenovirus A.*

Genus *Mastadenovirus* includes various viruses of mammals: viruses of cows, sheep, deer, pigs, dogs along with all human adenoviruses.

In humans, 57 adenoviruses are known, which are divided into 7 groups (A–G).

3. Structure

Adenoviruses are non-enveloped viruses, 80–90 nm in diameter. The icosahedral capsid of adenoviruses consists of hexones carrying group-specific and type-specific antigens and pentones containing mainly group-specific antigens at each apex. From each pentone there is a fiber with a head at the end (**Figure 1**).

Structural proteins of adenoviruses are designated by Roman numerals in descending order of molecular weight. The adenovirus capsid consists of seven structural proteins; three major capsid proteins hexon, fiber, and penton; and four minor proteins protein IIIa (pIIIa), VI, VIII, and protein IX [5]. Fibers provide binding to cellular receptors and participate in the discrimination of infected cells, causing inhibition of the synthesis of cellular macromolecules [6]. Soluble proteins of pentone cause a cytopathic effect similar to the action of infectious adenoviruses,

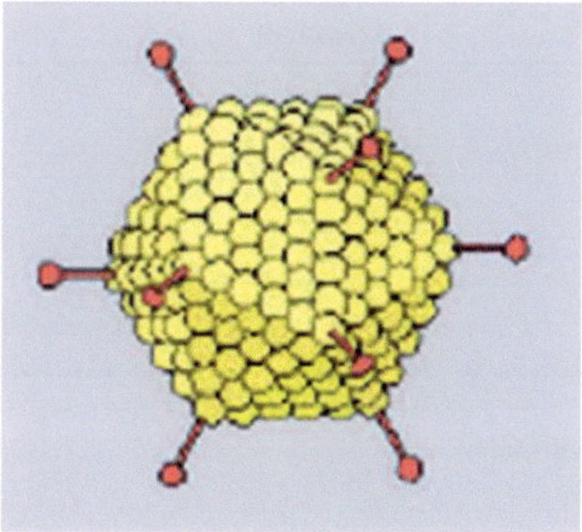

Figure 1. Adenovirus structure (from Linda Stannard, of the Department of Medical Microbiology, University of Cape Town).

but it manifests itself much faster (after 4–6 h). The pathogenetic significance of hexons is confirmed by the fact that antibodies against their epitopes demonstrated a neutralizing effect. They may be involved in the development of receptor-dependent endocytosis initiated by fibers, the main target for neutralizing antibodies [7].

A feature of adenoviral DNA is the presence of a terminal protein (TP), which is covalently linked to the 5′ end of each of the DNA strands. One of the possible functions of TP is the DNA attachment to the nuclear matrix after viral genome has entered the nucleus. Due to the interaction of TP, DNA is retained in the form of a ring structure, thus increasing the efficiency of transfection of adenoviral DNA isolated from virions [8]. The size of DNA is (20–25) 103 kDa, which corresponds to approximately 36,000 base pairs. This would be enough for about a dozen medium-sized proteins, but the information capacity of DNA is much more: adenoviruses synthesize about 40 proteins. This is achieved by reading information from both DNA strands and alternative splicing, which provides several types of mRNA based on the primary transcript of one gene. The 13 proteins are included in the mature virion, the rest belong to nonstructural components, functioning at the stage of intracellular reproduction of the virus [9].

4. Antigenic structure

Antigenic structure of *Mastadenovirus* is represented by three soluble antigens: the hexone A-antigen is common for all serotypes; pentone antigen (B-antigen) is a toxic antigen, it inhibits the action of interferon and increases the severity of associated respiratory infections; and fibril C-antigen is a type-specific which promotes the adsorption of adenoviruses on monkey or rat erythrocytes and causes their agglutination. Manifest forms of the disease cause epidemic serotypes (3, 4, 7, 14, 21) of subgroups B and E. Serotypes 1, 2, 5, and 6 subgroups C cause a latent current, contributing to the formation of chronic tonsillitis and adenoiditis. Clinical forms of adenovirus infection presented in **Table 1**.

Clinical manifestations	Serotype	Ref.
Acute respiratory diseases	1, 7, 14, 21	[10]
Viral pneumonia	1, 3, 4, 7, 55	[10–12]
Conjunctivitis	3, 7, 8, 10, 14, 19, 37	[13]
Gastroenteritis	9, 11, 31, 40, 41	[9, 14]
Meningoencephalitis	2, 6, 7, 12, 32	[15]

Table 1. Clinical forms of adenovirus infection.

The WHO have reported that data on the incidence and ramp prevalence of adenoviral infections are not exact, since in many cases, adenoviruses cause mild forms and therefore remain unregistered, according to general practitioners [16]. However, in recent years, adenoviruses caused widespread outbreaks in Asia [17–19] in which adenoviral infection was accompanied by the development of acute respiratory distress syndrome (ARDS) in Malaysia [18], China [16, 19], and South Korea [19].

5. Transmission

The source of infection is a sick person in the acute stage of the disease, convalescent, or a virus carrier. Pathogens are secreted with nasopharyngeal secretions, sputum, conjunctival discharge, feces, and urine (mainly in individuals with immunosuppression). The timing of isolation of pathogens from the upper respiratory tract reaches the 25th day of illness onset and more than 1.5 months with feces. Adenoviral infections are transmitted by airborne, introducing the virus to the conjunctiva and possibly by the fecal-oral route, thus affecting not only the respiratory tract but also other organs. The widespread disease is 5–10% of all viral diseases. Incidence is recorded throughout the year with a rise in the cold season. Both sporadic cases and epidemic outbreaks are observed. The most susceptible to infection are children from 6 months to 5 years, as well as military personnel. Particularly it has a high incidence in the newly formed groups of children and adults (in the first 2–3 months). In 95% of the adult population, antibodies to the most common serotypes of the virus can be detected in the serum [20].

6. Reproduction

The deproteinization of viruses entered the cell starts in the cytoplasm and ends in the nucleus, where DNA is released with a terminal protein attached to it. Transcription of the genome and replication of viral DNA occur in the nucleus with the help of cellular enzymes [21]. First, mRNAs are synthesized, which code for the synthesis of virus-specific enzymes and then also RNAs that carry information on the synthesis of capsid proteins and strands. The assembly of virus particles occurs in the nucleus, where crystal-like inclusions are formed. Several hundred viral particles are synthesized in each cell. The release of adenoviruses is accompanied by the destruction of the host cell. The cycle of reproduction of adenoviruses in the cell lasts 14–24 h.

Adenoviruses can multiply in different cells, including airway epithelial cells and lymphocytes. Virus-infected cells become targets for immunity effectors. However, adenoviruses impose such properties on cells that allow them to avoid destruction or at least reduce this possibility. Adenoviruses produce factors that block the synthesis and expression of HLA-I molecules on the cell surface and thereby inhibit the presentation of viral antigens attacked by CD8+T-lymphocytes. Cells infected with adenoviruses acquire increased resistance to interferons and TNF-a, a potent cytotoxic cytokine. In both cases, adenoviral proteins interfere with molecular mechanisms that determine antiviral effects. Among the products of early adenoviral genes are factors that delay the development of apoptosis. In general, this reflects a strategy aimed at improving the survival of infected cells and creating conditions for viral persistence [22].

7. Pathogenesis

From the pathogenetic point of view, adenoviruses damage the respiratory tract and belong to the "respiratory" viruses. However, adenoviruses also can cause lesions of the intestine and conjunctiva, as well as the central nervous system, bladder, and genitals. Adenoviruses multiply in the mucous membrane with a gradual, consistent involvement in the pathological process of the descending parts of the respiratory tract. Reproduction of adenoviruses also can occur in the intestinal tissue or lymph nodes which is accompanied by a multiple increase in lymph nodes. In addition to local changes, adenoviruses have a general toxic effect which appears as a fever and symptoms of general intoxication (weakness, lethargy, loss of appetite, headache, and nausea). The ability of adenoviruses to reproduce in the epithelial cells of the respiratory tract, conjunctiva, and intestine with the occurrence in some cases of hematogenous dissemination creates a wide range of clinical manifestations of this infection, including the appearance of generalized lymphadenopathy and widespread exanthema. In addition to adenoviruses in the genesis of acute pneumonia, the attachment of a secondary bacterial flora is important, which is facilitated by the suppression of the immune system [23–24].

8. Adenoviral latency

The term "adenoviruses" has an ecological coloration, reflecting the tendency to persistence in the lymphoid tissue, so that adenoviruses can be isolated from the tonsils, adenoids, appendix, and lymph nodes of practically healthy people.

After acute infections, many serotypes are not eliminated from the body for a long time. Most of the latent viruses belong to subgroups B2 and C. They can persist for years in the lymphoid tissue of the pharyngeal ring and, apparently, other localizations (e.g., in the mesenteric lymph nodes). Once in the intestine, adenoviruses asymptomatically replicate in the epithelial cells and/or Peyer's patches, periodically evolving from feces.

Latency creates the possibility of endogenous recurrences of acute infection and chronic hyperplasia (in fact, chronic inflammation) of infected lymphoid tissue. The possibility of activating a viral infection in the tonsils (chronic tonsillitis), mesenteric lymph nodes, and appendix is not excluded. Adenoviruses can be activated on the background of immunosuppressive therapy in AIDS patients. A number of new serotypes (serotypes 43–47 of subgroup D) were first isolated from AIDS patients (from feces), allowing to believe that a persistent infection creates favorable conditions for the evolution of adenoviruses [25].

The mechanism of persistence of adenoviruses in the lymphoid tissue still remains unclear. Most likely, this is due to the low content of sensitive (permissive) cells and very slow replication of the virus in lymphocytes, that is, with severely limited productive infection. It is significant that in the experiments of W. Rowe et al., who discovered adenoviruses in 1953, it took several weeks for the degeneration of a culture of adenoid tissue associated with the reproduction of latent adenoviruses. The possibility of integrative virogeny with partial expression of the viral genome is not excluded. In approximately 50% of the tonsils in adults, it was possible to detect adenoviral antigens in the absence of an infectious virus.

9. Antiadenovirals

As mentioned above, adenoviruses are unsusceptible to interferons. The acyclic nucleoside phosphonates HPMPC (cidofovir) [26] exhibit antiadenoviral activity. However, these substances are toxic, so they are applied locally in the form of ointments or by injection directly into the affected organ. The target of HPMPC is adenoviral DNA polymerase. The mechanism of the antiviral effect of this compound is based on its phosphorylation by cell kinases and in the synthesis of DNA by competition with conventional nucleosides. At consecutive inclusion in a thread of two molecules of HPMPC, elongation is blocked. The targeted effect of NM on infected cells is provided by a higher affinity of HPMPC diphosphate for viral DNA polymerase than for host DNA polymerase.

10. Adenovirus vaccines

Adenovirus infection is perhaps the only respiratory disease other than influenza, against which specific prevention methods have been developed. Military contingents in the United States since 1971 are vaccinated with live oral adenovirus vaccine against serotypes 7 and 4 which were isolated in the 1950s [27]. Live oral adenovirus vaccine has proven to be safe and highly effective in numerous clinical trials, as well as in clinical observations of acute respiratory infections among US military personnel. However, live oral adenoviral vaccine types 4 and 7 are approved only for use in military teams for adults 17–50 years old. To avoid the virus being thrown into the upper respiratory tract, it is recommended to swallow the tablets whole, without chewing. Adenovirus is extremely stable under natural conditions, and there is the possibility of being released into the environment through excreta. Therefore, a vaccine containing live strains of adenovirus is not recommended for use in children or the general population. Despite the long-term stability of the adenovirus genome, which is confirmed by the efficacy of ongoing vaccination among the US military, random mutations or homologous recombination events, which can lead to changes in the antigenicity of adenovirus, are not excluded. In addition, over time, the epidemic types of adenovirus have changed, and in recent years, highly pathogenic serotypes 14 and 55 have been distributed throughout the world. Finally, the circulation of the types of adenoviruses can vary geographically; for example, in China, the most common types associated with acute respiratory infections (ARI) are types 3, 7, and 55 [28]. Therefore, attention should be paid to developing new adenoviral vaccines based on currently circulating adenovirus strains.

11. The mutagenic effect

Representatives of the genus *Mastadenovirus* served as objects of research in the field of molecular biology. The intrigue around the pathogenetic function of adenoviruses aggravated after their tumorigenic properties, the ability to induce malignant tumors in animals (newborn hamsters), was recorded. In 1962, Trentin et al. described the first case of the induction of a malignant tumor in animals by the human pathogenic virus—adenovirus-12, which caused tumors in rodents [29]. The oncogenic potential of adenoviruses has served as a stimulus for their careful study, which has proved useful for studying the mechanisms of viral infection and molecular processes in eukaryotic cells. On the model of adenovirus infection, splicing, adenylation, and capturing of matrix RNA, sequence and expression regulation of viral genes, their integration with cellular chromosomes, etc. were first studied with human tumors. In a series of works [30, 31], it was shown that high-oncogenic type 12 adenoviruses and type 18 adenoviruses induce chromosomal aberrations in nonpermissive cells for them. The mutagenic effect at the chromosomal level has been demonstrated for human type 2 adenoviruses and type 5 in rat cells [32], as well as for bovine adenovirus type 3 in Chinese hamster cells [33]. In humans, despite intensive research, the association of malignant tumors with adenoviruses has not been identified.

12. The use of adenoviruses as vectors for gene therapy

The promise of using adenoviruses as vectors is due to the fact that a relatively large fragment can be inserted into their linear DNA. With the advent of second-generation vectors, it became possible to embed foreign DNA sequences up to 35 kb in the adenovirus genome, while maintaining only inverted repeats and packaging site. In addition, adenovirus receptors (e.g., termination of the fibers) can be genetically modified in such a way as to increase the tropism of the virus in relation to the tumor tissue. As a product of the transgene, which allows to destroy a tumor, you can use the herpes virus thymidine kinase (family *Herpesviridae*) or the chicken anemia virus apoptin (family *Circoviridae*). In the first case, the patient is prescribed of acyclovir; in the second case, the tumor is destroyed as a result of vector-induced apoptosis. Unfortunately, genetically engineered constructions based on adenoviral vectors have not yet found clinical application, since due to the use of multiple mechanisms by the virus for penetration into target cells, it is not possible to achieve selective delivery of vectors to cancer cells [34, 35].

13. Adenovirus vectors vaccines

Adenovirus vectors first appear in the 1980s. They have received great attention as gene delivery systems for vaccine antigens and were extensively tested in several preclinical and clinical studies. Adenovirus-based vector vaccines have been developed and are being studied against a variety of infectious diseases, including influenza, measles, hepatitis B, rabies, anthrax, Ebola, severe acute respiratory syndrome (SARS), human immunodeficiency virus 1 (HIV-1), malaria, tuberculosis [36–40]. A study of an oral tableted vaccine based on a non-replicative recombinant adenovirus-5 serotype carries DNA that encodes the hemagglutinin of A/California/04/2009 (H1N1) pdm with an adjuvant added in the form of double-stranded RNA (dsRNA) [41] showed that after a single immunization, significant systemic and local immunity occurs. Seroconversions of anti-hemagglutinating antibodies to A/H1N1 pandemic influenza viruses were detected in 92% of participants [41].

The advantages of adenoviral vectors are that they efficiently transfer genes to both dividing and nondividing cells, do not integrate into the genome, and provide high titers of recombinant virus and high expression levels of the introduced genes. Although currently used adenoviral vectors may cause nonspecific inflammation and antiviral cellular response [42]; this problem still needs to be solved.

14. Conclusions

The study of human and animal adenoviruses allows not only to explore the molecular relationship with the host organism but also to solve specific problems of medicine. In the practical field, on the one hand, new antiviral vaccines and chemotherapy drugs have been developed, implemented, or are awaiting introduction; on the other hand, there is a perspective on the therapeutic use of viruses, such as oncolytic viruses or viral vectors. These and other problems associated with adenoviruses and adenoviral infections will be covered in more detail in subsequent chapters of this book.

Author details

Yulia Desheva[1,2]*

*Address all correspondence to: desheva@mail.ru

1 Federal State Budgetary Scientific Institution "Institute of Experimental Medicine", Saint Petersburg, Russian Federation

2 Saint Petersburg State University, Saint Petersburg, Russian Federation

References

[1] Huebner RJ, Rowe WP, Ward TG, Parrott RH, Bell JA. Adenoidal-pharyngeal-conjunctival agents: A newly recognized group of common viruses of the respiratory system. New England Journal of Medicine. 1954;**251**(27):1077-1086

[2] Robbins FC, Enders JF, Weller TH. Cytopathogenic effect of poliomyelitis viruses in vitro on human embryonic tissues. Proceedings of the Society for Experimental Biology and Medicine. 1950;**75**(2):370-374

[3] Hilleman MR, Werner JH. Recovery of new agent from patients with acute respiratory illness. Proceedings of the Society for Experimental Biology and Medicine. 1954;**85**(1):183-188

[4] Enders JF, Bell JA, Dingle JH, Francis T Jr, Hilleman MR, Huebner RJ, et al. "Adenoviruses": Group name proposed for new respiratory-tract viruses. Science. 1956;**124**:119-120

[5] Nemerow GR, Pache L, Reddy V, Stewart PL. Insights into adenovirus host cell interactions from structural studies. Virology. 2009;**384**(2):380-388

[6] Reddy VS, Natchiar SK, Stewart PL, Nemerow GR. Crystal structure of human adenovirus at 3.5 A resolution. Science. 2010;**329**:1071-1075

[7] Vellinga J, van der Heijdt S, Hoeben RC. The adenovirus capsid: Major progress in minor proteins. The Journal of General Virology. 2005;**86**(Pt 6):1581-1588

[8] Liu H, Jin L, Koh SB, Atanasov I, Schein S, Wu L, et al. Atomic structure of human adenovirus by cryo-EM reveals interactions among protein networks. Science. 2010;**329**:1038-1043

[9] Qiu FZ, Shen XX, Li GX, Zhao L, Chen C, Duan SX, et al. Adenovirus associated with acute diarrhea: A case-control study. BMC Infectious Diseases. 2018;**18**(1):450

[10] Centers for Disease Control and Prevention (CDC). Acute respiratory disease associated with adenovirus serotype 14—Four states, 2006-2007. Morbidity and Mortality Weekly Report. 2007;**56**:1181

[11] Zhang SY, Luo YP, Huang DD, Fan H, Lu QB, Wo Y, et al. Fatal pneumonia cases caused by human adenovirus 55 in immunocompetent adults. Infectious Diseases. 2016;**48**: 40-47. DOI: 10.3109/23744235.2015.1055585

[12] Jain S, Williams DJ, Arnold SR, Ampofo K, Bramley AM, Reed C, et al. Community acquired pneumonia requiring hospitalization among U.S. children. The New England Journal of Medicine. 2015;**372**:835-845. DOI: 10.1056/NEJMoa1405870

[13] Singh MP, Ram J, Kumar A, Rungta T, Gupta A, Khurana J, et al. Molecular epidemiology of circulating human adenovirus types in acute conjunctivitis cases in Chandigarh, North India. Indian Journal of Medical Microbiology. 2018;**36**(1):113

[14] Reis TAV, Assis ASF, Valle DAD, Barletta VH, Carvalho IPD, Rose TL, et al. The role of human adenoviruses type 41 in acute diarrheal disease in Minas Gerais after rotavirus vaccination. Brazilian Journal of Microbiology. 2016;**47**(1):243-250

[15] Similä S, Jouppila R, Salmi A, Pohjonen R. Encephaloningitis in children associated with an adenovirus type 7 epidemic. Acta Paediatrica Scandinavica. 1970;**59**:310-316. DOI: 10.1111/j.1651-2227.1970.tb09009.x

[16] World Health Organization (WHO). Viruses. In: Pond K, editor. Water Recreation and Disease. Plausibility of Associated Infections: Acute Effects, Sequelae and Mortality. London, UK: IWA Publishing. Available from: http://www.who.int/water_sanitation_health/bathing/recreadischap6.pdf [Accessed: December 14, 2017]

[17] Cheng J, Qi X, Chen D, Xu X, Wang G, Dai Y, et al. Epidemiology and transmission characteristics of human adenovirus type 7 caused acute respiratory disease outbreak in military trainees in East China. American Journal of Translational Research. 2016;**8**:2331

[18] Yusof MA, Rashid TR, Thayan R, Othman KA, Hasan NA, Adnan N, et al. Human adenovirus type 7 outbreak in Police Training Center, Malaysia, 2011. Emerging Infectious Diseases. 2012;**18**:852-854

[19] Jeon K, Kang CI, Yoon CH, Lee DJ, Kim CH, Chung YS, et al. High isolation rate of adenovirus serotype 7 from South Korean military recruits with mild acute respiratory disease. European Journal of Clinical Microbiology & Infectious Diseases. 2007;**26**:481-483

[20] Kajon A, Lynch J. Adenovirus: Epidemiology, Global Spread of Novel Serotypes, and Advances in Treatment and Prevention. Seminars in Respiratory and Critical Care Medicine. 2016;**37**(04):586-602

[21] Smith JG, Wiethoff CM, Stewart PL, Nemerow GR. Adenovirus. In: Cell Entry by Non-Enveloped Viruses. Berlin, Heidelberg: Springer; 2010. pp. 195-224

[22] Cook J, Radke J. Mechanisms of pathogenesis of emerging adenoviruses. F1000 Research. 2017;**6**:90

[23] Lenaerts L, De Clercq E, Naesens L. Clinical features and treatment of adenovirus infections. Reviews in Medical Virology. 2008;**18**(6):357-374

[24] Khanal S, Ghimire P, Dhamoon AS. The repertoire of adenovirus in human disease: The innocuous to the deadly. Biomedicine. 2018;**6**(1):30

[25] Morillo SG, Luchs A, Cilli A, Carmona RC, Neme SN, Timenetsky MC. Rotavirus genotype G4P [8] and enteric adenovirus in HIV-positive patients with and without diarrhea in São Paulo state, Brazil. Transactions of the Royal Society of Tropical Medicine and Hygiene. 2010;**104**(2):165-167

[26] De Clercq E. Clinical potential of the acyclic nucleoside phosphonates cidofovir, adefovir, and tenofovir in treatment of DNA virus and retrovirus infections. Clinical Microbiology Reviews. 2003;**16**(4):569-596

[27] Hilleman MR, Greenberg CJH, Warfield MS, Anderson SA, Glabere RR. Second field evaluation of bivalent types 4 and 7 adenovirus vaccine. AMA Archives of Internal Medicine. 1958;**102**(3):428-436

[28] Chen S, Tian X. Vaccine development for human mastadenovirus. Journal of thoracic disease. 2018;**10**(Suppl 19):S2280

[29] Trentin JJ, Yabe Y, Taylor G. The quest for human cancer viruses: A new approach to an old problem reveals cancer induction in hamsters by human adenovirus. Science. 1962;**137**(3533):835-841

[30] Stich HF, Hammerberg O, Casto B. The combined effect of chemical mutagen and virus on DNA repair, chromosome aberrations and neoplastic transformation. Canadian Journal of Genetics and Cytology. 1972;**14**(4):911-917

[31] McDougall JK. Adenovirus-induced chromosome aberrations in human cells. Journal of General Virology. 1971;**12**(1):43-51

[32] Paraskeva C, Roberts C, Biggs P, Gallimore PH. Human adenovirus type 2 but not adenovirus type 12 is mutagenic at the hypoxanthine phosphoribosyltransferase locus of cloned rat liver epithelial cells. Journal of Virology. 1983;**46**(1):131-136

[33] Lukash LL, Buzhievskaya TI, Varshaver NB, Shapiro NI. Oncogenic adenovirus as mutagen for Chinese hamster cells in vitro. Somatic Cell Genetics. 1982;**7**(2):133-146

[34] Eisenberger A, Elliott BM, Kaufman HL. Viral vaccines for cancer immunotherapy. Hematology/Oncology Clinics of North America. 2006;**20**:661-687. PubMed: 16762729

[35] Verma IM, Naldini L, Kafri T, Miyoshi H, Takahashi M, Blömer U, et al. Gene therapy: Promises, problems and prospects. Nature. 1997;**389**(6648):239-242. DOI: 10.1007/978-3-642-56947-0_13

[36] Gao W, Soloff AC, Lu X, et al. Protection of mice and poultry from lethal H5N1 avian influenza virus through adenovirus-based immunization. Journal of Virology. 2006;**80**:1959-1964

[37] Fooks AR, Schadeck E, Liebert UG, et al. High-level expression of the measles virus nucleocapsid protein by using a replication-deficient adenovirus vector: Induction of an MHC-1-restricted CTL response and protection in a murine model. Virology. 1995;**210**:456-465

[38] Hashimoto M, Boyer JL, Hackett NR, et al. Induction of protective immunity to anthrax lethal toxin with a nonhuman primate adenovirus-based vaccine in the presence of preexisting anti-human adenovirus immunity. Infection and Immunity. 2005;**73**:6885-6891

[39] Lubeck MD, Natuk RJ, Chengalvala M, et al. Immunogenicity of recombinant adenovirus-human immunodeficiency virus vaccines in chimpanzees following intranasal administration. AIDS Research and Human Retroviruses. 1994;**10**:1443-1449. PubMed: 7888199

[40] Chengalvala MV, Bhat BM, Bhat R, et al. Immunogenicity of high expression adenovirus-hepatitis B virus recombinant vaccines in dogs. The Journal of General Virology. 1994;**75**(Pt 1):125-131

[41] Liebowitz D, Lindbloom JD, Brandl JR, Garg SJ, Tucker SN. High titre neutralizing antibodies to influenza after oral tablet immunisation: A phase 1, randomised, placebo-controlled trial. The Lancet Infectious Diseases. 2015;15:1041-1048. DOI: 10.1016/S1473-3099(15)00266-2

[42] Agnandji S, Huttner A, Zinser M, Njuguna P, Dahlke C, Fernandes J, et al. Phase 1 Trials of rVSV Ebola Vaccine in Africa and Europe. New England Journal of Medicine. 2016;**374**(17):1647-1660

Chapter 2

Chemotherapy of Adenovirus Infections

Angel S. Galabov

Additional information is available at the end of the chapter

http://dx.doi.org/10.5772/intechopen.79160

Abstract

Adenoviruses occupy a substantial place as causative agents of seasonal respiratory infections, the most characteristic and severe being epidemic keratoconjunctivitis (EKC). Moreover, adenovirus infections are very characteristic with their severe course in persons with impaired immune system. The absence of specific anti-adenovirals is the major problem, and the development of compounds effective against adenoviruses is a principal task. This chapter embraces the results of studies on search for antivirals with anti-adenovirus activity, nucleoside/nucleotide analogues and nonnucleoside compounds. Ganciclovir and cidofovir demonstrated effects against adenovirus serotypes *in vitro* and in animal ocular infection models. Cidofovir applied alone or in combination with cyclosporine manifested therapeutic effects on patients with EKC in a controlled clinical study. We characterized abitylguanide as anti-adenovirus agent in broad-scale investigations, including cell culture experiments, and in two double-blind trials with very beneficial results.

Keywords: human adenoviruses, antivirals, nucleosides/nucleotides, nonnucleosides, mode of action, clinical trials

1. Role of adenoviruses in human pathology

Adenoviruses (AdVs) [1] occupy a significant place in the human pathology [2]. It was established that 30–70% of the human population in Europe and North America shows a seroprevalence to AdVs (cit. in [3]). These viruses are causative agents of wide range of human diseases, showing varying tissue tropism. They are generally middle and self-limiting. Among them, large place is occupied by acute respiratory tract diseases, especially in children [4]. Conjunctivitis is very often registered in these infections. More severe course manifests viral gastroenteritis, especially in infants, as well as hemorrhagic cystitis, and in rare cases, hepatitis, myocarditis, meningoencephalitis or nephritis [5, 6]. AdVs are characteristic with their severe course in persons with hereditary decreased or impaired immune system: (1) hereditary immunodeficiency; (2) in persons transplant recipients treated with immunosuppressive agents; (3) AIDS. In such patients, the abovementioned clinical manifestations are particularly prone to disseminated disease frequently show a fatal outcome, in children mortality rate attains 83% [7–9]. EKC is another serious and very frequent AdV induced disease, extremely often with social importance [10–14].

The major problem of AdVs infections is the absence of chemotherapeutic agents not only for the clinical practice, but even the absence at strong anti-adenovirals in experimental research. This is pointed in all manuals of virology considering AdVs and AdV-induced infections, in the review articles and even in all encyclopedic sources. Evidently, development of an effective antiviral treatment is a principal task.

This chapter presents a concentrated view on the investigations of experimental chemotherapy of AdV infections and results of their clinical application (**Table 1**).

Compound name	Chemical structure	Target	Way of administration and clinical trial
Cidofovir	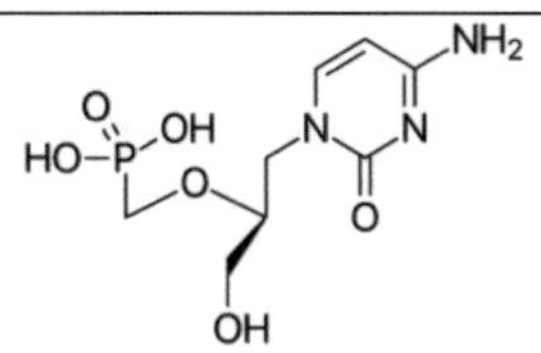	AdV DNA polymerase	*Local application in EKC patients* *Oral application in immunocompromised patients* VISTID Controlled clinical pilot study of combination with cyclosporine
Brincyclovir	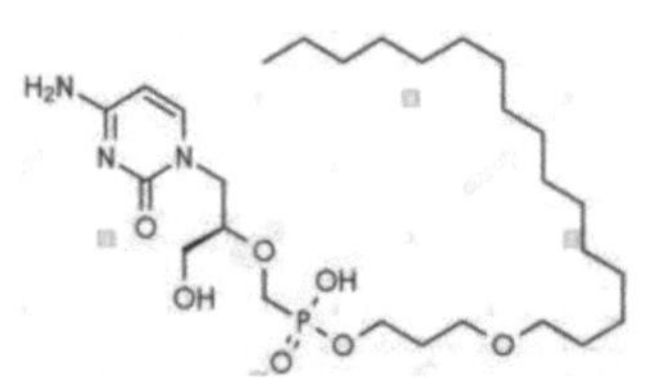	AdV DNA polymerase	*Oral application* CHIMERIX Phase II clinical trial
(S)-HPMPA	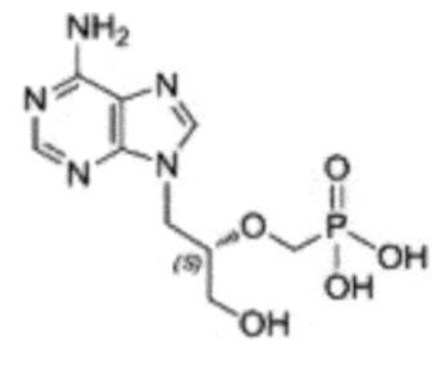	AdV DNA chain elongation	
(S)-HPMPO-DAPy	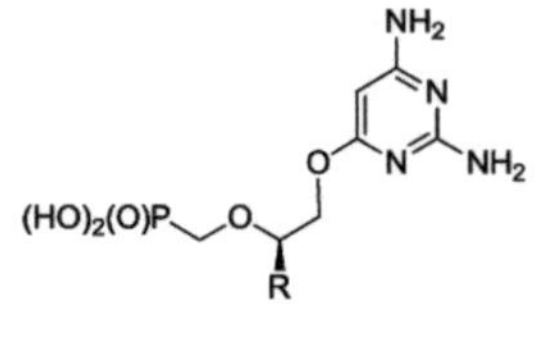	AdV DNA polymerase	

http://dx.doi.org/10.5772/intechopen.79160

Compound name	**Chemical structure**	**Target**	**Way of administration and clinical trial**
Ganciclovir	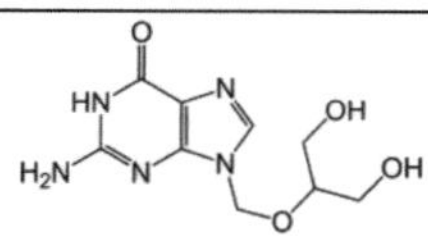	AdV DNA synthesis, inhibition of late genes expression	*Local application* ZIRGAN Three double-blind trials: two in US and one in Germany (40 placebo and 40 ZIRGAN treated each) GCV in combination with the microbicide povidon-iodine
Zalcitabine	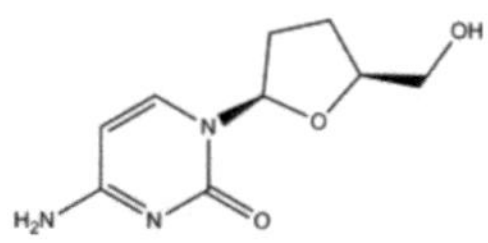		
Alovudine	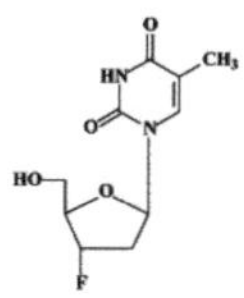	AdV DNA polymerase	
Trifluridine	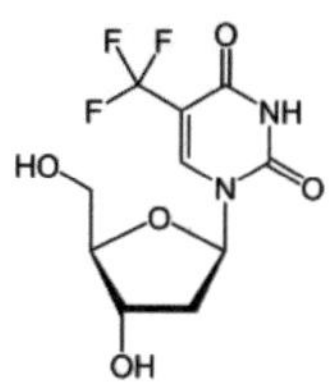	DNA synthesis	*Local application*
Vidarabine	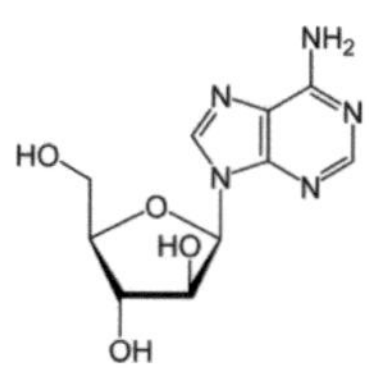	DNA synthesis	
Ribavirin	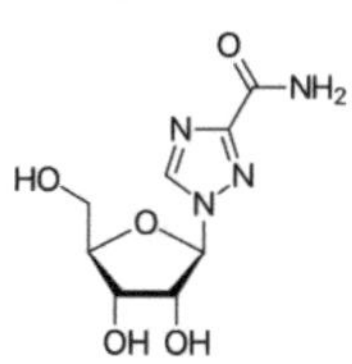	• Inhibition of inosin-5-MP dehydrogenese (decreased GTP pools) • Viral polymerases inhibition (RNA capping activity of viral transcripts) • Lethal mutagenesis of viral RNA genomes ("error catastrophe")	*General application in immunocompromized patients*
Abitylguanide-HCl	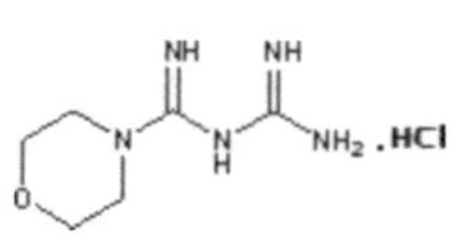	Ligand of AdV capsid proteins	*Local application in EKC patients* ADENOSTATIN COLLYRIUM Two placebo-controlled randomised trials on 349 EKC patients (151 and 198, respectively) carried out in 1973/74; three placebo-controlled trials in latest 80th years

Compound name	Chemical structure	Target	Way of administration and clinical trial
15D8	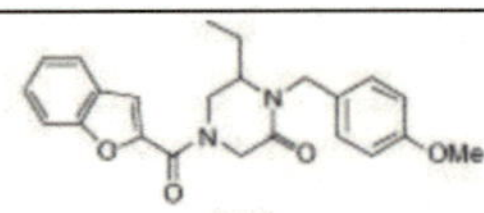 15D8	Ligand of precursors of terminal protein pTP, AdV DNA polymerase and DNA-binding protein	
Digoxin and Digitoxin	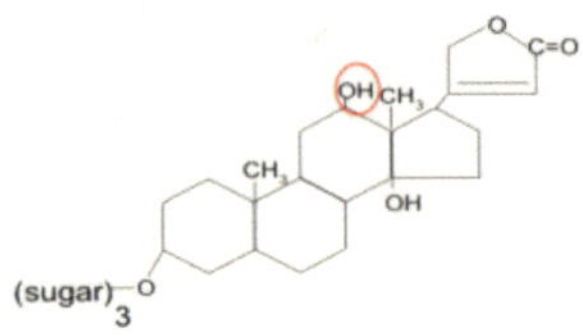	Alter RNA splicing EIA activity gene early in AdV infection (late proteins E4 and F6 and the major late capsid hexon protein are compromised)	*General application(?)*
	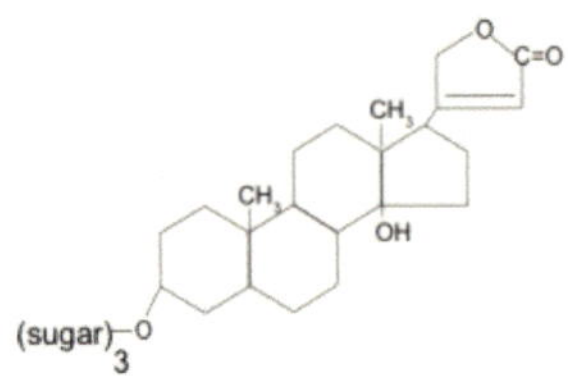		
2,4,5,7-Tetranitro-9-fluorenone	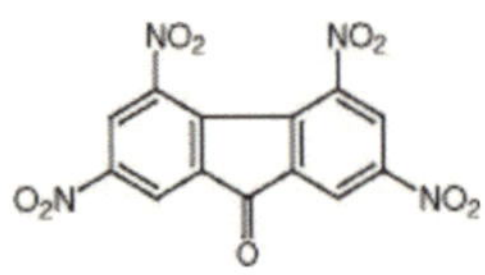	Inhibits AdV cysteine protease (indispensable for the production of infectious AdV virions)	
4-{4-[Benzyl(methyl) amino]-2-hydroxybenzoyl} benzoic acid	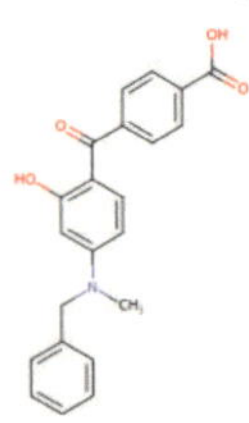		

Table 1. Compounds manifesting activity against human adenoviruses.

2. Nucleoside/nucleotide analogues

2.1. Cidofovir [(S)-HPMPC; (S)-1-(3-hydroxy-2-phosphonomethoxypropyl)cytosine; VISTID]

The antiviral effect of cidofovir is based on its transformation in the infected cells in di- and triphosphate and such way becoming an alternative substrate of the AdV DNA polymerase possessing higher affinity compared to cellular DNA polymerases (I, II and III) [15]. Sequence changes in a conserved region of the AdV DNA polymerase were established at cidofovir-resistant AdV mutants [16].

Cidofovir IC_{50} values in cell culture testing versus broad spectrum of AdVs are within 0.8–17 μg/ml [17]. In in vivo testing in ocular infection with AdV (type C) of New Zealand rabbits and cotton rats the compound treatment is efficacious when administered as 0.5–1% eye drops

[18–20]. Romanowski and Gordon [21] found efficacy of topical 0.5% cidofovir on several human adenoviruses (AdV1, AdV5 and AdV6) in the New Zealand rabbit ocular model. AdV type B and type C-induced pneumonia registered in mice and in cotton rats [22] could be used for in vivo treatment with antivirals.

2.2. Brincyclovir (BCV; hexadecyloxypropyl-cidofovir; CMX001)

This compound is a lipidic conjugate of cidofovir. It prevents AdV induced mortality in a permissive, immunosuppressed animal model [23].

2.3. (S)-HPMPA [(S)-9-(3-hydroxy-2-phosphonomethoxypropyl)adenine

Its anti-AdV effect has the same mechanism as cidofovir—inhibition of the AdV DNA chain elongation [24].

2.4. USC-187 (alkyl tyrosinamide-ester prodrug of HPMPA)

This compound proved active against multiple AdV serotypes in vitro and was effective versus AdV-C6 in hamsters immunosuppressed by cyclophosphamide. Administered orally USC-187 prevented or significantly decreased mortality, virus titers and liver pathology up to 4 days post AdV i.v. challenge. Applied in a respiratory AdV-C6 challenge model USC-187 manifested symptoms of toxicity [25].

2.5. (S)-HPMPO-DAPy [2,4-diamino-6-[3-hydroxy-2-(phosphonomethoxy)-propoxy] pyrimidine)

The compound anti-AdV activity registered was slightly inferior than that of cidofovir and HPMPA [26].

2.6. (S)-2242 [2-amino-7-(1,3-dihydroxy-2-propoxymethyl)purine

In vitro manifested a marked activity on AdV replication with selectivity index exceeding cidofovir [26].

2.7. Ganciclovir (GCV)

This compound is known as a drug approved for the treatment of herpes infection (cytomegalovirus infection especially) was reported to be effective against human AdVs in vitro [27]. In cell culture GCV inhibits AdV5 replication and expression of late genes [28]. These authors established a marked effect of 3% GCV in cotton rat eyes, on replication and pathology of this virus [28] Ying et al. [29] tested the GCV administered locally for prophylactic or therapeutic effect in immunosuppressed (by cyclophosphamide) Syrian hamsters intravenously infected with human AdV5 and was established that the compound suppresses AdV5 replication in the liver and AdC5-induced pathology of infected hamsters thus mitigated the consequences of the AdV infection. It was showed that GCV inhibits AdV5 DNA synthesis and late gene expression. The slight increase in GCV phosphorylation in AdV infected cells established by

Ying et al. [29] could be a result of slightly elevated cellular thymidine kinase activity, higher in testing in vivo. These authors hypothesize the direct inhibition of the AdV DNA polymerase as a possible mechanism of GCV suppressive effect on AdV DNA synthesis.

2.8. Zalcitabine (2′3′-dideoxycytidine, ddC)

This anti-HIV compound possesses a marked anti-AdV activity, even stronger than cidofovir in ocular AdV infections in laboratory animals [30].

2.9. Alovudine

Alovudine (FddT) manifest in vitro (IC_{50} = 0.2–0.7 μg/ml) and in vivo (mouse model of AdV pneumonia) anti-AdV activity of the order of that of cidofovir [26, 31].

2.10. Trifluridine (3FT) and vidarabine (Vira-A)

Anti-AdV activity of these anti-HSV compounds is moderate, and the current data on their testing are controversial [32–34].

2.11. Ribavirin (1-ß-D-ribofuranosyl-1,2,4-triazole-3-carboxamide)

This triazole nucleoside was described initially by Sidwell et al. [35]. Numerous studies were carried out on the mode of action of this compound manifesting activity toward large spectrum of viruses (predominantly RNA containing) belonging to different taxonomic groups. However, there are no data about the mechanism of anti-AdV effect of ribavirin. Several different mechanisms were formulated about the antiviral effects of this compound: (1) decreased levels of intracellular guanosine triphosphate pools based on inhibition of cellular inosine-5-monophosphate dehydrogenase; (2) inhibition of viral polymerases; (3) inhibition of RNA capping activity of viral transcripts; (4) lethal mutagenesis of the viral RNA genomes, also termed "error catastrophe," based on the induction of increased viral mutation rate over the critical error rate (especially expressed on enterovirus replication) via incorporation of the compound into newly synthesized genomes; (5) immunomodulatory role particularly on adaptive immune responses—the compound is inducer of the helper-T-cell type 1 cytokine response, but also a suppressor of the type 2 cytokine phenotype. Data about anti-AdV effect in cell culture experiments are very controversial. It was established that its activity is limited to AdVs of group C and strongly cell culture-dependent [36]. However, the plasma concentrations reached by ribavirin are 10 times below the required IC_{50} value [36, 37].

3. Nonnucleoside compounds

3.1. Abitylguanide

N′N′-anhydro-bis(β-hydroxy-ethyl)biguanide-HCl (abitylguanide) suppressed markedly the replication of a large spectrum of human AdVs, both standard laboratory strains and strains isolated from epidemic keratoconjunctivitis patients. The magnitude of inhibitory effect varied from 1.5 to 3.8 logs. A marked correlation was established between the value of inhibitory effect and belonging of tested AdVs to various subgroups, the strongest activity being found toward viruses of subgroup C (Rosen's subgroup III). The compound susceptible period of AdV 5 replication included the total growth cycle, but is especially pronounced during the exponential phase. This was established through timing-of-addition study in primary cell cultures of human embryo kidney cells. Electron microscopy study of AdV5 morphogenesis contributed substantially to the clearing-up of the mechanism of anti-AdV action of abitylguanide. It was registered that: (1) the compound decreased about 10-fold the percentage of cells in which mature or empty virions with the characteristic nuclear localization were observed; (2) a complete absence of paracrystals; (3) the number of cells with virus particles arranged in crystals in the nucleoplasm was strongly decreased [38].

The absence of crystalline inclusions established electron microscopically in our study correlated with the established pronounced decrease of infectivity and/or lower yields of viral progeny which is in line with the meaning [39] that the protein crystals might be involved in at late steps of the virus life cycle ensuring correct capsid assembly, virus maturation and infectivity. Discussing the mode of action of abitylguanide on AdV 5 replication it has to stress on the full coincidence of the data obtained electron microscopically with the results of the timing-of-addition study demonstrating the highly-pronounced compound-sensitivity of the virus growth in the exponential phase. On the basis of the mentioned data, abitylguanide can be considered a ligand of AdV capsid protein(s) [38].

3.2. Trisubstituted piperazin-2-one derivative 15D8

This piperazinone is a result of a large screening of low-molecular substances, embracing chemical libraries of in total more than 25,000 compounds. A prospective selection of the compounds was based on protein-protein and protein-DNA interaction [40, 41]. The derivative 15D8 showed substantial anti-AdV activity (AdV 5 and AdV16 models) in dose-dependent manner at high MOI (15,000 vp/cell) with little or absent cytotoxicity at low micromolar concentration. The compound selectively inhibits AdV DNA replication in the nucleus. It is possible 15D8 to interact with viral proteins essential for DNA, including precursor of the terminal protein (pTP), AdV DNA polymerase or the DNA-binding protein (DBP). 15D8 could be considered as a potential candidate for the development of a new class of antiviral compounds to treat AdV infections [42].

3.3. Cardiotonic steroids—digoxin and digitoxin

Very surprising recently (2017), the cardiotonic steroids entered in the scope of the struggle with AdV infections. As a theoretic prerequisite of their effects were the data on dependents of AdV on the host pre-RNA splicing machinery for expression of its complete genome. On such base modulators of RNA splicing as digoxin and digitoxin could be considered as antivirals versus human AdVs. Grosso et al. [3] proved that both drugs reduced of a series of AdVs of four different species (A to D) by 2–3 logs. This is a result of affecting several steps needed for AdV genome replication (late proteins E4 or f6 and the major late capsid hexon protein is compromised). The authors proved that these two drugs altered EiA RNA splicing early in infection and partially blocked the translation from 12S and 13S to 9S RNA at later stages of viral replication. By blocking AdV replication at one or more steps beyond the onset of E1A expression and before genome replication, digoxin and digitoxin manifest very prospective potential as antivirals for the treatment of serious AdV infections.

Convallatoxin, a synthetic cardiotonic steroid, manifested a stronger activity versus AdV5 when compared with digoxin and digitoxin. In general, these three substances alter the cascade of AdV gene expression—an effect starting after initiation of early gene expression attaining a blocking of AdV DNA replication and of viral structural proteins. These findings open a novel approach of treating AdV infections and guide the development of novel antiviral therapies.

3.4. $NMSO_3$ (sulfated sialic acid derivative)

This compound inhibits selectively cellular binding sites (sialic acid-containing receptors) of several AdV serotypes. It was established that ADVs possess this cellular tropism. $NMSO_3$ could be used for the topical treatment of ocular AdV infections [43, 44].

3.5. 2,4,5,7-tetranitro-9-fluorenone

This substance markedly inhibits the AdV cystein protease, indispensable for the production of infectious AdV virions [45, 46].

3.6. ARD-209 (ADENOVIR)

As a result of the studies carried out by G. Wadell, N. Arnberg and others in University of Umea in Sweden (2014–2017), O. Sterner and U. Ellervik (University of Lund, Sweden) synthesized the substance APD-209, announced as ADENOVIR (Pharma). It was noted that this substance with unknown structure for the publicity is considered as a new solution for the treatment of viral eye infections [47].

Other Swedish authors [48] reported also that **analogues of 2-[2-(benzoylamino)benzoyl amino]benzoic acid** possess inhibitory effect on adenovirus replication.

3.7. Short interfering RNAs

In the search for discovery of efficient anti-AdV agents it was investigated the probable impact of silencing of a set of early, middle and late viral genes on the replication of AdV5 in vitro [49]. It was established that AdV replication was inhibited by siRNAs directed against AdV E1A, DNA polymerase, preterminal protein (pTR), IVa2, hexon, and protease genes. Besides, silencing of the early and middle genes was more effective in inhibiting AdV replication than the silencing of late genes, especially sharply manifested with effect on siRNA of the DNA polymerase gene. Besides, it was found that reducing the viral genome copy numbers (AdV DNA) is a more promising strategy than the reducing the number of proteins necessary for capsid formation.

3.8. Other compounds inhibiting AdV replication in vitro

Cyclic D,L-α-peptides, cycloferon, lactoferrin nitric oxide, doxovir, heterocyclic Schiff bases of aminohydroxyguanidine tosylate, PGD peptidomimetic molecules, some medical plant substances (ref. in [17]).

4. Clinical trials

Prophylaxis with effective antivirals versus AdV EKC would be particularly useful for preventing AdV transmission to the second eye as well as to the contact persons.

Cidofovir eye drops 1% prevent severe corneal opacities in EKC patients, dose-dependent local toxicity at frequent administration been registered [50, 51]. This anomalous nucleoside at concentration of 1% applied topically in a combination with 1% cyclosporine demonstrated therapeutic effect on patients with EKC in a controlled clinical pilot study [51]. No placebo-controlled randomized trials have been carried out on immunocompromised patients.

Controversial results were obtained with cidofovir the treatment of AdV infections in children undergoing bone marrow or stem cell transplantation [52]. Evidence was accumulated that as earlier AdV infection is detected for starting cidofovir treatment the better curative results were registered. Although cidofovir exhibits antiviral activity versus all AdV species, it possesses low oral bioavailability and significant toxicity (tubular necrosis) and does not confer long-term protection [53].

However, the lipid conjugate of cidofovir, brincyclovir (BCV; hexadecyloxy propyl-cidofovir; CMX001) is currently in Phase II clinical trial [25, 54], unfortunately manifesting a significant toxicity to the kidney and gastrointestinal tract.

The topical ganciclovir application against EKC [55] merits special attention. In a published clinical study, treatment with 0.15% GCV ophthalmic gel improved outcome of AdV conjunctivitis [56]. Three other clinical trials evaluating 0.15% GCV ophthalmic gel were organized, two of them finished: (1) Efficacy and Safety of GV 550 in Acute Adenovirus Keratoconjunctivitis (Clinical Trials.gov. trial NCT01156025) and (2) Efficacy and Safety of GV 550 in Acute Adenovirus Keratoconjunctivitis (a Clinicaltrialsregister.eu trial). Both included placebo and treatment groups on 40 persons each (in fact Phase II of clinical trial). In the third trial, Clinical Trials.gov trial NCT1533480 (A Placebo Controlled Comparison of Topical ZIRGAN Versus Genteal for the treatment of Adenovirus Conjunctivitis) which is currently in course, GCV is administered topically as 0.15% gel (ZIRGAN®) compared with 0.3% Hypromellose gel (Genteal gel®), serving as placebo (Phase IV).

On the base of the promising results with povidone-iodine (PVP-I), a microbicidal agent possessing also virucidal properties [57], topical ganciclovir and PVP-I combination drops have shown the most recent potential, but both therapeutics need to be investigated in larger scale studies [58].

The experimental data in vivo are in favor of GCV to be considered as an option for the treatment of AdV infections in immunocompromised patients.

Ribavirin efficacy for the treatment of AdV infections was very controversial [36, 59].

N-chlorotaurine (a week oxidant) manifested effectivity in phase II clinical trials with viral conjunctivitis [60].

During a severe outbreak of EKC caused by AdV 8 in 1972–1973 in Bulgaria abitylguanide was tested in two double-blind, randomized trials, carried out on total 349 patients (trial 1–151 patients; trial 2–198 patients) with virologically confirmed diagnosis.

Abitylguanide was applied as 1% eye drops (in saline). In each of the two trials, patients were divided in three groups: group I placebo (patients with symptomatic treatment), group II -abitylguanide 1% + symptomatic treatment, group III—patients treated with abitylguanide 1% only. Curative effect of abitylguanide was almost identical in group II and III in both trials. Moreover, the drug had a preventive effect on infection of the second eye. The abitylguanide treatment exerted a marked curative effect on the severity and duration of the disease: (1) more than twofold decrease in both trials in the number of patients with EKC form associated with keratitis; (2) five- and sixfold decrease in trial 1 an trial 2, respectively, in the incidence of severe keratitis; (3) two- and fivefold decrease in trial 1 and trial 2, respectively, in the number of patients with impaired vision; (4) twofold decrease of the healing time [61–63]. Effectivity toward AdV-induced EKC of the drug applied topically was confirmed in series of placebo controlled trials (not following the double-blind scheme) in three other ophthalmic clinics in the country, carried out in the second half of the 1980s. Pencheva et al. [64] in the Varna Medical Faculty registered marked decrease of the patients number with keratitis, two-fold shortening of the healing time in abitylguanide treated patients, affection of the second eye—80% in the placebo group, and 22.6% in abitylguanide treated group.

In preliminary carried out study abitylguanide manifested a very high local tolerance (1, 2 and 3% eye drops in saline) tested in 21 volunteers (in rabbits—till 20% eye drop).

The pronounced effect of abitylguanide [65] in abovementioned trials on EKC patients served for the development and implementation in pharmaceutical industry of preparative ADENOSTATIN COLLYRIUM® (Pharmachim Ltd., Sofia) which clinical use marked favorable estimation by ophthalmologists in this country (tested in Japan, as well).

The clinical use of cardiotonic steroids as anti-AdV agents needs special consideration. As digitoxin has been associated with toxicity, the use of cardiac glycosides as antivirals would be short term, in contrast to the chronic use in patients with heart diseases. Having in mind that anti-AdV agents have to be used for the treatment of severe respiratory and disseminated diseases, these drugs seem more attractive as potential agents for the topical treatment of EKC and even for the prophylaxis of persons contact to EKC patients [3].

There is no doubt that chemotherapy of AdV infections occupies leading position as a tool for anti-etiological treatment. Therefore, the development of effective anti-AdV agents is especially a big task of the scientists and clinicians. The above-presented panorama of antivirals versus AdVs and AdV-caused infections shows that a lot of work is done for the realization of this problem. The author would like to mention the main directions that determined the development of antivirals and their implementation in the clinical practice: (1) discovery of the targets in virus growth cycle for chemotherapeutic attacks—a lot of "wide" places could be pointed in the AdVs; (2) attainment in the organic chemistry—modeling of new effective molecules with anti-AdV effects among anomalous nucleosides, end especially of non-nucleoside compounds ligands of AdV proteins; (3) development of adequate methodology for antiviral testing starting from the initial in vitro screening, and application of purified AdV structural and eventually nonstructural proteins as cell-free systems (approach contributed substantially for the successful development of anti-hepatitis C drugs; as concerns the in vivo testing, in the last years, several very convenient and adequate models were described and successfully used (more precisely ocular AdV infections in laboratory animals); (4) application of methods for express diagnostic of ADV infections, in order earliest start of the respective treatment with anti-AdV chemotherapeutic agents. More detailed consideration of this topic was presented by Kaufman [66] and Luchs [67].

Author details

Angel S. Galabov

Address all correspondence to: galabov@microbio.bas.bg

The Stephan Angeloff Institute of Microbiology, Bulgarian Academy of Sciences, Sofia, Bulgaria

References

[1] Robinson CM, Singh G, Lee JY, Dehghan S, Rajaiya J, Liu EB, Yousuf MA, Betensky RA, Jonas MNS, Dyer DW, Seto D, Chodosh J. Molecular evolution of human adenoviruses. Scientific Reports. 2013;**3**:1812

[2] Ghebremedhin B. Human adenovirus: Viral pathogen with increasing importance. European Journal of Microbiology & Immunology (Bp). 2014;**4**:26-33

[3] Grosso F, Stoilov P, Brown M, Cochrane A. Suppression of adenovirus replication bu cardiotonic steroids. Journal of Virology. 2016;**91**(3):e01623-e01616

[4] Brandt CD, Kim HW, Vargosko AJ, Jeffries BC, Arrobio JO, Rindge B, Parrot RH, Chanock RM. Infections in 18,000infacnts ands children in a control study of respiratory tract disease. I. Adenovirus pathogenicity in relation to serologic type and illness syndrome. American Journal of Epidemiology. 1969;**90**:484-500

[5] Strausberger R, Harel L, Levy Y, Amir J. A syndrome of transient encephalopathy associated with adenovirus infection. Pediatrics. 2001;**107**:E69

[6] Chuang YY, Chiu CH, Wong KS, Huang JG, Chang LY, Lin TY. Severe adenovirus infection in children. Journal of Microbiology, Immunology, and Infection. 2003;**36**:37-40

[7] Munoz FM, Piedra PA, Demmler JG. Dissminated adenovirus disease in immunicompromised and immunocompetent children. Clinical Infectious Diseases. 1998;**27**:1194-1200

[8] Echavarria M. Adenoviruses in immunocompromised hosts. Clinical Microbiology Reviews. 2008;**21**:704-715

[9] Matthes-Martin S, Boztug H, Lion T. Diagnosis and treatment of adenovirus infections in immunocompromised patients. Expert Review of Anti-Infective Therapy. 2013;**11**:1017-1028

[10] Butt AL, Chodosh J. Adenoviral keratoconjunctiovitis in a tertiary care eye clinic. Cornea. 2006;**25**:199-202

[11] Jin X, Ishiko H, Nata NT. Molecular epidemiology of adenoviral conjunctivitis in Hanoi, Vietnam. American Journal of Ophthalmology. 2006;**113**:1064-1066

[12] Bialansiewicz A. Adenoviral keratoconjunctivitis. Sultan Qaboos University Medical Journal. 2007;**7**:15-23

[13] Sambursky RP, Fram N, Cohen EJ. The prevalence of adenoviral conjunctivitis at the Wille eye hospital emergency room. Optometry. 2007;**78**:236-239

[14] Jhanji V, Chan TCY, Li EYM, Agarwal K, Vajpayee RB. Adenoviral keratokonjunctivitis. Survey in Ophthalmology. 2015;**60**:435-443

[15] De Clercq E. Clinical potential of the acyclic nucleoside phosphonates cidofovir, adefovir, and tenofovir in treatment of DNA virus and retrovirus infections. Clinical Microbiology Reviews. 2003;**16**:569-596

[16] Kinchington PR, Araullo-Cruz T, Vergnes JP, Yates K, Gordon YJ. Sequence changes in the human adenovirus type 5 DNA polymerase associated with resistance to the broad spectrum antiviral cidofovir. Antiviral Research. 2002;**56**:73-84

[17] Lenaerts L, Naesens L. Antiviral therapy for adenovirus infections. Antiviral Research. 2006;**71**:172-180

[18] Gordon YJ, Romkanowski E, Araullo-Cruz T. An ocular model of adenovirus type 5 infection in the NZ rabbit. Investigative Ophthalmology & Visual Science. 1992;**33**:574-580

[19] De Oliveira CB, Stevenson D, La Bree L, McDonnell PJ, Throusdale MD. Evaluation of C idofovir (HPMPC, GS-504) against adenovirus type 5 infection in vitro and in a New Zealand rabbit ocular model. Antiviral Research. 1996;**31**:165-172

[20] Kaneko H, Mori S, Suzuki O, Iida T, Shigeta S, Abe M, Ohno S, Aoki K, Suzutani T. The cotton rat model for adenovirus ocular infection: Antiviral activity of sidofovir. Antiviral Research. 2004;**61**:63-55

[21] Romanowski EG, Yates KA, Gordon YJ. The in vitro and in vivo evaluation of ddC as a topical antiviral for ocular adenovirus infections. Investigative Ophthalmology & Visual Science. 2009;**50**:5295-5299

[22] Kajon AE, Gigliotti AP, Harrod KS. Acute inflammatory response and remodeling of airway epithelium after subspecies B1 human adenovirus infection of the mouse lower respiratory tract. Journal of Medical Virology. 2003;**71**:233-244

[23] Toth K, Spencer JF, DDhar D, Sagarttz JE, Buller RM, Paiter GR, Wold WS. Hexadecyloxypropyl-cidofovir, CMX001, prevents adenovirus-induced mortality in a permissive immunosuppressed animal model. Proceedings of the National Academy of Sciences of the United States of America. 2008;**105**:7293-7297

[24] Mul YM, van Miltenburg RT, De Clercq E, van der Vilet PC. Mechanism of inhibition of adenovirus DNA replication by the acyclic nucleoside triphosphate analogue

(S)-HPMPApp: Influence of the adenovirus DNA binding protein. Nucleic Acids Research. 1989;**17**:8917-8929

[25] Toth K. In: Abstracts of 30th International Conference on Antiviral Research, Atlanta, Antiviral Res. 145; 2017

[26] Naesens L, Lenaerts L, Andrei G, Snoeck R, Van Beers D, Holy A, Balzarini J, De Clercq E. Antiadenovirus actrivities of several classes of nucleoside and nucleotide analogues. Antimicrobial Agents and Chemotherapy. 2005;**49**:1010-1016

[27] Kinchington PR, Romanowski EG, Gordon YJ. Prospects for adenovirus antivirals. The Journal of Antimicrobial Chemotherapy. 2005;**55**:424-429

[28] Trousdale MD, Goldschmidt PL, Nobrega R. Activity of ganciclovir against human adenovirus type-5 infection in cell culture and cotton rat eyes. Cornea. 1994;**13**:435-439

[29] Ying B, Tollefson AE, Spencer JF, Balakrishnan L, Dewhurst S, Capella C, Buller RML, Toth K, Wold WSM. Ganciclovir inhibits human adenovirus replication and pathogenicity in permissive immunosuppressive Syrian hamsters. Antimicrobial Agents and Chemotherapy. 2014;**58**:7171-7181

[30] Romanowski EG, Gordon YI. Efficacy of topical cidofovir on multiple adenoviral serotypes in the New Zealand rabbit ocular model. Investigative Ophthalmology & Visual Science. 2000;**41**:460-463

[31] Mentel R, Wegner U. Evaluation of the efficacy of 2′, 3′-dideoxycytidine against adenovirus infection in a mouse pneumonia model. Antiviral Research. 2000;**47**:79-87

[32] Waring GOE, Laibson PR, Satz JE, Joseph NH. Use of vidarabine in epidemic keratoconjunctivitis due to adenoviruses 3, 7, 8, and 19. American Journal of Ophthalmology. 1976;**82**:781-785

[33] Lennetts DA, Eiferman RA. Inhibition of adenovirus replication in vitro by trifluridine. Archives of Ophthalmology. 1978;**96**:1662-1663

[34] Ward JB, Siojo LG, Wallere SG. A prospective, masked clinical trial of trifluridine, dexamethazone, and artificial tears in the treatment of epidemic keratoconjunctivitis. Cornea. 1993;**12**:216-221

[35] Sidwell RW, Robins RK, Hilyard IW. Ribavirin: An antiviral agent. Pharmacology & Therapeutics. 1979;**6**:123-146

[36] Morfin F, Dupuis-Girod S, Mundweiler S, Falcon D, Carrington D, Sedlacek P, Bierings M, Cetkovski P, Kroes AC, van Tol MJ, Thouvenot D. In vitro susceptibility of adenovirus to antiviral drugs is species-dependent. Antiviral Therapy. 2005;**10**:225-229

[37] Morfin F, Dupuis-Girod S, Frobert E, Mundweiler S, Carrington D, Sedlacek P, Bierings M, Cetkovski P, Kroes AC, van Tol MJ, Thouvenot D. Differential susceptibility of adenovirus clinical isolates to cidofovir and ribavirin is not related to species alone. Antiviral Therapy. 2009;**14**:55-61

[38] Galabov AS, Vassileva V, Karabasheva V. Inhibitory effect of abitylguanide on adenovirus replication. Drug Research. 2017;**67**(10):583-590 2017-02-1359/24.5.2017/MPS

[39] Franqueville L, Henning P, Magnusson M, Vigne E, Schoehn G, Blair-Zajdel ME, Habib N, Lindholm L, Blair GE, Hong SS, Boulanger P. Protein crystals in adenovirus type 5-infected cells: Requirements for intranuclear crystallogenesis, structural and functional analysis. PLoS One. 2008;**3**(8):e2894. DOI: 10.1371/journal.pone.0002894

[40] Boger DL, Desharnais J, Capps K. Solution-phase combinatorial libraries: Modulating cellular signaling by targeting protein-protein or protein-DNA interactions. Angewandte Chemie International Edition England. 2003;**42**:4138-4176

[41] Whitby LR, Boger DL. Comprehensive peptidomimetic libraries targeting protein-protein interactions. Accounts of Chemical Research. 2012;**45**:1698-1709

[42] Sanchez-Cespedes J, Moyer CL, Whitby LR, Boger DL, Nemerow GR. Inhibition of adenovirus replication by a trisubstituted piperazin-2-one derivative. Antiviral Research. 2014;**108**:65-73

[43] Arnberg N, Kidd AH, Edlund K, Olfat F, Wadell G. Initial interactions of subgenus D adenoviruses with A549 cellular receptrors: Sialic acid versus alpha(γ) integrins. Journal of Virology. 2000;**74**:7691-7693

[44] Kaneko H, Kato K, Mori S, Shigeta S. Antiviraql activity of $NMSO_3$ against adenovirus in vitro. Antiviral Research. 2001;**52**:281-288

[45] Pang YP, Xu K, Kollmeyer TM, Perola E, McGrath WJ, Green DT, Mandel WF. Discovery of a new inhibitor lead of adednovirus proteinase: Steps toward selective, irreversible inhibitors of cysteine proteinases. FEBS Letters. 2001;**502**:93-97

[46] Mandel WF, Baniecki ML, McGrath WJ. Specific interections of the adenovirus proteinase with the viral DNA, an 11-amino-acid viral peptide, and the cellulaqr protein actin. Cellular and Molecular Life Sciences. 2003;**60**:2347-2355

[47] Greber UF, Arnberg N, Wadel G, et al. Adenoviruses: From pathogens to therapeutics: A report on the 10th international adenovirus meeting. Cellular Microbiology. 2013;**15**:16-23

[48] Öberg CT, Strand M, Andersson EK, Eglund K, Tran NPN, Mei Y-F, Wadel G, Elofsson M. Synthesis, biological evaluation and structure-activity relationship of 2-[2-(benzoylamino)benzoylamino] benzoic acid analogues as inhibitors of adenovirus replication. Journal of Medicinal Chemistry. 2012;**55**:3170-3181

[49] Kneidinger D, Ibrisimovic M, Lion T, Klein R. Inhiobition of adenovirus multiplication by short interfering RNAs directly on indirectly targeting the viral DNA replication machinery. Antiviral Research. 2012;**94**:195-207

[50] Hillenkamp J, Reinhard T, Ross RS, Bohringer D, Catsburg O, Roggendorf M, De Clercq E, Godenhardt E, Sundmacher R. Topical tereatment of acute adenoviral keratoconjunctivitis with 0.2% cidofovir and 0.1% cyclosporine: A controlled clinical pilot study. Archives of Ophthalmology. 2001;**119**:1487-1491

[51] Hillenkamp J, Reinhard T, Ross RS, Bohringer D, Catsburg O, Roggendorf M, De Clercq E, Godenhardt E, Sundmacher R. The effects of cidofovir 1% with and without

of cyclosporin a 1% as a topical treatment of acute adenoviral keratoconjunctivitis: A controlled clinical pilot study. Ophthalmology. 2002;**109**:845-850

[52] Ljungman P, Ribaud P, Eyrich M, Matthesw-Martin S, Eins4ele H, Bleakley M, Machaczka M, Bierings M, Bosi A, Grateos N, Cordonnier C. Cidofovir for adenovirus ibfections after allogeneic hematopoietic stem cell trnspolantation: A survey by the infectious Dioseases Woreking Party of the European Group for blood and marrow Transplatation. Bone Marrow Transplantation. 2003;**31**:481-486

[53] Lindemans CA, Leen AM, Boelens JJ. How I treat adenovirus in hematopoietic stem cell transplant recipients. Blood. 2010;**116**:5476-5485

[54] Paolino K, Sande J, Perez E, Loechelt B, Jantaussch B, Painer W, Anderson M, Tippin T, Lanier ER, Fry T, DeBiasi RL. Eradication of disseminated adenovirus infection in a pediatric hematopoetic stem cell transplantation recipient using the novel antiviral agent CMX001. Journal of Clinical Virology. 2011;**50**:167-170

[55] Tabbara KF, Jarade EF. Ganciclovir effects in adenoviral keratoconjunctivitis. [abstracts 3111-B253] ARVO 42: S579; 2001

[56] Yabiku ST, Yabiku MM, Bottos KM, Araujo AL, Freitas D, Belfort R Jr. Ganciclovir 0.15% ophthalmic gel in a treatment of adenovirus keratoconjunctivitis. Arquivos Brasileiros De Oftalmologia. 2011;**74**:417-421

[57] Isenberg SJ, Apt L, Valenton M. A controlled trial of povidone-iodine to treat infectious conjunctivitis in children. American Journal of Ophthalmology. 2002;**134**:681-688

[58] Pihos AM. Epidemic keratoconjunctivitis: A review of current concepts in management. Journal of Optometry. 2013;**6**:69-74

[59] La Rosa AM, Champlin RE, Mirza N, Gajewski J, Giralt S, Rolston KV, Raad I, Jacobson K, Kontoyiannis D, Elting L, Whimbey E. Adenovirus infectionms in adult recipients of blood and marrow transplants. Clinical Infectious Diseases. 2001;**32**:871-876

[60] Teuchner B, Nagl M, Schidlbauer A, Ishiko H, Dragosits E, Ulmer H, Aoki K, Ohno S, Mizuki N, Gottardi W, Larcher C. Tolerability and efficacy of N-chlorotaurine in epidemic keratoconjunctivitis - a double-blind, randomized, phase-2 clinical trial. Journal of Ocular Pharmacology and Therapeutics. 2005;**21**:157-165

[61] Wassileva PI, Galabov AS. Uber die Behandlung der epidemischen Keratokonjunktivitis mit ABOB Klinische und Laboruntersuchungen. Klinische Monatsblätter für Augenheilkunde. 1975;**166**:77-83

[62] Vassileva P, Galabov AS. Combined therapy of epidemic keratoconjunctivitis. In: Bialasiewicz A, editor. Update of Infectious Diseases of the Eye. New York: Springer Verlag; 1993. pp. 307-314

[63] Arshinkov N, Galabov A. Treatment of epidemic keratoconjunctivitis with ABOB (in Bulgarian, abstr. In English). Epidemiologiya, mikrobiologiya, infektsiozni bolesti. 1976;**13**:200-205

[64] Pencheva D, Galabov A, Tuncheva K, Stojanova R. The clinical effects of adenostatin in the therapy of epidemic keratoconjunctivitis. Bolletino di oculistica. 1987;**66**(5):135-136

[65] Galabov BS, Lozanova HS, Galabov AS. Method for obtaining of N'N'anhydrobis (β-hydroxyethyl) biguanide hydrochloride. Bulgarian Iinvention Certificate No. 41554, reg. No. 74436/8.04.1986; transformed in Patent No. 341/24.02.1994; 1986

[66] Kaufman HE. Adenovirus advances: New diagnostic and therapeutic options. Current Opinion in Ophthalmology. 2011;**22**:290-293

[67] Luchs JI. Cutting-edge diagnostic technologies foster patients' Confidence. Cataract & Refrective Surgery Today June 2014; 2014. pp. 79-82

Chapter 3

Adenovirus as Tools in Animal Health

José M. Rojas, Noemí Sevilla and Verónica Martín

Additional information is available at the end of the chapter

http://dx.doi.org/10.5772/intechopen.79132

Abstract

Adenoviruses have long been identified as good candidates for use as viral vectors in gene therapy and as vaccines. These viruses can infect multiple cell types, while in division or in quiescence, and are relatively easy to manipulate so that parts of their genome can be replaced with exogenous genes. Progressive safety improvements in replication-deficient adenoviral vectors have been achieved with the second and third generation, and ending with the gutless adenoviral vectors. Adenoviral vectors are immunogenic and can act as adjuvants. Nonetheless, the potency of human recombinant adenoviral vaccines was below expectations in clinical trials mainly because of the pre-existing adenoviral immunity found in the general population. This drawback can however become advantageous in animal health, as no previous immunity to human adenoviral vectors exists in animals. Other viral vectors viruses are used as vaccine, but adenoviruses remain the most employed and promising recombinant vector in veterinary medicine. In this chapter, we review the generation of adenoviral vectors, the immune response they trigger, and their advantages and disadvantages for veterinary use in terms of safety and efficacy. This chapter also describes how recombinant adenoviral vectors can be integrated as tools for vaccination and immunomodulation in veterinary medicine.

Keywords: adenovirus vectors, vaccines, animal health, immune response

1. Introduction to adenoviral vectors

1.1. Adenovirus

Adenoviruses (Ad) are large (90-100 nm), nonenveloped, not segmented, and linear double-stranded DNA viruses belonging to the viral family *Adenoviridae* that infect a broad range of vertebrate hosts, from fish to humans. They replicate in the nucleus of the infected cells. These viruses have an icosahedral nucleocapsid consisting of three major proteins called hexon (or protein II), penton base (or protein III), and a nodulated fiber (or protein IV) together with a number of other minor proteins, VI, VIII, IX, IIIa, and Iva2. This capsid contains 26–48 Kbp double-stranded DNA genome (**Figure 1A**), which has a terminal protein (TP) attached to one of its ends. They were first isolated in 1953 from a culture of human adenoid cells, hence their name [1]. Of the more than 100 Ad described since then, 57 infect humans causing conjunctivitis, hemorrhagic cystitis, gastroenteritis, and respiratory diseases. The *Adenoviridae* family contains five genera based on DNA composition and host species: *Aviadenovirus, Atadenovirus, Mastadenovirus, Siadenovirus,* and *Ichtadenovirus* [2]. Within the genera, the viruses are grouped into species, and named from the host followed by letters of the alphabet. For example, the human adenoviruses (HuAd) are classified within the *Mastadenovirus* genus and divided into seven subgroups, from A to G [3, 4]. Classification questions remain, however, unresolved for many nonhuman adenoviruses.

1.2. Adenoviral vectors

Viral vectors are modified viruses used to introduce exogenous DNA into host cells, and their construction uses similar principles. Virus functions can be divided into elements that act in *cis* such as the origins of replication or the encapsidation sequence that must be found in the genome of the viral vector, or act in *trans* such as structural proteins and/or envelope or the machinery necessary for viral replication that do not need to be encoded by the viral genome itself. These *trans* elements can be supplied by stably transfected cells (packaging cells), or through transient transfections with plasmids or helper virus. The general method for viral vector construction consists in substituting the *trans* elements, essential for replication, by the gene of interest. The most popular technique developed for constructing replication-defective (RD) recombinant adenoviral vectors is that described by Dr. F. Graham and known as the "two-plasmid method" (available in commercial kits) [5]. Nonreplicative (defective) particles thus obtained maintain the infectivity of the parental virus, but are unable to produce new infective viral particles, and possess the ability to transfer the therapeutic gene material introduced into their genome. The viruses most commonly used as vectors are poxviruses, retroviruses, and Ad.

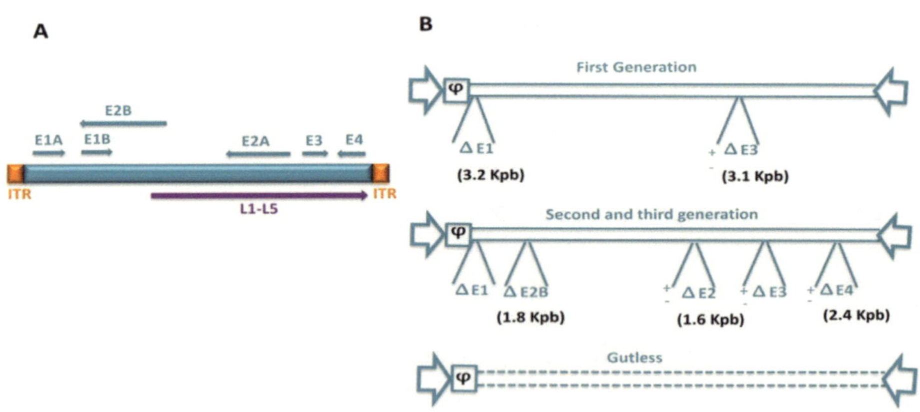

Figure 1. (A) Schematic representation of the adenoviral genome organization. E, early genes; L, late genes; and ITR, inverted terminal repeat sequences. (B) Diagram of the evolution of the different adenoviral vectors. Deletions (Δ) from different areas of the adenoviral genome have improved these vectors in terms of capacity to house an exogenous gene and in terms of safety, avoiding reversions. ψ, cis packaging signal.

Ad possess most of the qualities required to be a successful viral vector. They infect a large variety of mitotic and postmitotic cells replicating episomally without chromosomal integration, thus reducing the risk of insertion mutations and oncogenesis. They have high cloning capacity, high transduction efficiency, and high transgene expression. They are relatively easy to prepare and purify, which permits the obtention of high viral titer with low toxicity. HuAd serotypes 2 and 5 are the best characterized and most used for creating recombinant vectors [6]. The RDAd used as vectors can be divided into three classes, schematized in **Figure 1B** [7, 8], according to the size of the deletions made in their genome, which directly impacts on the size of the exogenous DNA they can harbor.

From a safety point of view, it is preferable to work with replication-defective (RD) Ad [9], and this chapter will mainly focus on RDAd. There are nonetheless several studies that use replication-competent (RC) Ad in veterinary vaccination, for instance to improve mucosal immunity or override maternal-derived immunity [10, 11].

2. Immunogenicity of adenoviral vectors

RDAd vectors induce humoral, cellular, and mucosal protective immune responses in a variety of animal models [12]. They are particularly suited to produce potent cellular immune response to the encoded antigens [13]. Vector innate immunogenicity and antigen expression affect and shape the adaptive immune response triggered by RDAd infection.

Innate immune responses are essential for triggering an effective adaptive response. RDAd activate nucleotide-binding oligomerization domain-like receptor (NLR) and toll-like receptor (TLR) signaling pathways and induce several cytokines such as IL-1, IL-12, IL-6, TNF, and interferon (IFN)-α. Myeloid differentiation protein-88 (MyD88) signaling contributes to the induction of RDAd adaptive immune response since systemic and mucosal immunity was reduced in MyD88-deficient mice after RDAd vaccination [14]. $CD8^+$-T cell responses elicited after RDAd vaccination are, however, not dependent on TLRs or IL1-R family member since T-cell responses are not significantly diminished in mice lacking different TLRs, IL-1R, or IL-8R [15]. Type I IFN production and signaling probably participate to transgene immunity. Type I IFN levels correlate with transgene neutralizing antibody titers [16] and IFN-β promoter stimulator-1 (IPS-1) and type I IFN signaling are required for the induction of antigen-specific $CD8^+$-T cells in the gut mucosal compartment [17]. Besides TLRs, cells detect cytosolic viral DNA through NLRs, which are at the core of the inflammasome that triggers inflammatory responses producing IL-1β, IL-18, and IL-6 (reviewed in [18]). NF-κB-dependent inflammatory gene expression (IL-1β, IL-6, and MIP-1β) was significantly reduced in NALP3-deficient mice after RDAd inoculation [19], indicating that the NALP3 inflammasome mediates the innate immune response to RDAd.

The magnitude and quality of the T cell immune response elicited by RDAd is influenced *in vivo* by the vector cellular tropism, which alters the source of cytokines and chemokines produced during vaccination. After intravenous inoculation, Kupffer cells in liver [20] and macrophages in the marginal zone of the spleen [21] are infected by RDAd, whereas after subcutaneous or intramuscular inoculation (the most commonly used vaccination routes), $CD11c^+$

dendritic cells (DCs) are transduced in the draining lymph node. The $CD11c^{+}CD8^{-}B220^{-}$ compartment showed enhanced RDAd uptake and transgene expression [22], but in spite of being less frequently transduced, the $CD11c^{+}CD8^{+}B220^{-}$ DC subset was more potent at inducing T cell proliferation against the transgene. $CD11c^{+}$ DCs are, therefore, critical for eliciting T cell responses against RDAd-encoded transgenes.

High transgene antigen-specific responses after infection with Ad serotypes, such as HuAd5, are associated with high transgene expression levels *in vivo* [23]. The amount and duration of the antigen expression is thus one of the most relevant parameters that shape the immune response induced by RDAd. In mice, HuAd5 and chimpanzee-derived ChAd3 produce high and persistent antigen expression with low innate immunity activation resulting in strong T cell response induction, whereas RDAds that express less antigen and trigger a robust innate immunity are less potent inducers of T cell responses [23].

Pre-existing vector-specific humoral and cellular immunity limits the duration of transgene expression and is one of the main problems for RDAd uses as vaccines [24]. Vector-specific neutralizing-antibodies suppress the immunogenicity of adenoviral vector vaccines [25]. Although neutralizing antibodies are serotype specific and mainly directed against the hypervariable loops of the viral hexon, non-neutralizing antibodies to more conserved regions of the adenoviral particle cross-react between Ad serotypes [26]. Passive antibody transfer from RDAd-immunized animals to naïve animals demonstrated that adeno-specific neutralizing antibodies reduced the induction of transgene-specific $CD8^{+}$ T cells after homologous challenge. Nonetheless, these neutralizing antibodies change the fate of the $CD8^{+}$ T cells and promote their transition into the memory cell pool [27]. This could be highly relevant for vaccine design, since enhanced $CD8^{+}$ cell expansion to the transgene can be detected when boost inoculation was given with a heterologous RDAd.

It, thus, appears that the balance between immunity to the vector and the transgene defines successful RDAd vaccination strategies. Recognition of the vector is necessary for Ad adjuvancy to take place, while high transgene expression and immunogenicity are also required to drive the immune response toward the antigen of interest.

3. Recombinant adenoviral vectors in veterinary medicine

3.1. Considerations for veterinary vaccines and adenoviral vector vaccines

The use of vaccines to fight animal diseases is one of the most efficient strategies of preventive medicine regarding cost-effect ratio. It helps reduce disease, minimizes long-term healthcare costs, and ultimately reduces inequity in health [28]. Maladies such as rinder pest have been eradicated thanks to vaccine campaigns. Multiple parameters need to be considered for a potential vaccine to become successful, such as its efficacy, safety and immunogenicity, and the possibility of large-scale production at low cost while maintaining genetic stability. Ideally, a vaccine should also be single dose and provide long-term systemic and mucosal immunity [29].

In veterinary medicine, adenoviral vectors that express immunogenic pathogen proteins have been used as vaccine to activate a protective immune response to the pathogen [30, 31]. The use of HuAd5, most commonly used in human trials, in animal health can be advantageous, as no previous immunity to this adenoviral vector should exist in animals. Recombinant Ad strongly activate the immune system [32] and generate immunity toward both the vector and the expressed transgene. These strong humoral and cell-mediated antigen-specific responses [12, 13] are a prerequisite for a good vaccine candidate that can even preclude for adjuvant need. But it may also present a problem, since immunity to the vector could be generated in vaccinated animals, which would limit efficacy if a second immunization was needed. Several approaches can be undertaken to solve this problem, from using a single inoculation to induce protection, to using heterologous prime-boost systems or using different Ad serotypes for consecutive inoculations [33].

RDAd recombinant vectors can be produced in large scale with a high titer [34] and lyophilized, or produced in thermostabilized forms [35] so that they can be easily stored and transported, even in conditions in which the maintenance of a cold chain can be problematic as in case of distribution to remote locations in hot climate countries. For veterinary medicine, vaccines need to be particularly inexpensive. As part of the One Health strategy, vaccination also offers the added benefit of limiting antibiotic use in animal production, either through direct vaccination effects or by limiting viral diseases that can lead to opportunistic bacterial infections.

3.2. Adenoviral vectors as DIVA vaccines

Most veterinary vaccines do not allow infected-recovered animals to be distinguished from vaccinated animals, the so-called differentiating infected from vaccinated animals (DIVA) approach. DIVA vaccines can be used as control tools for disease outbreaks, limiting animal culling in the eradication process. They, thus, have a great economic importance as they facilitate animal health status monitoring and grant disease-free status more quickly to countries affected by an outbreak. RDAd expressing antigenic proteins are suitable DIVA vaccines as vaccinated animals that only respond to proteins encoded by the vaccine can be differentiated from infected animals that also respond to viral proteins not encoded by the RDAd vaccine. An adenovirus-based vaccine was shown to be successful as foot and mouth disease (FMDV) DIVA vaccine [36]. RDHuAd5 that express peste des petits ruminants virus (PPRV)-F or -H proteins are another example of DIVA veterinary vaccines [37–39]. While vaccinated animals developed antibodies against F and H, infected animals also developed antibodies against N, and due to validated commercially available tests for anti-N and anti-H antibodies, infected animals could be differentiated from vaccinated animals. RDAd-based vaccines appear, thus, particularly suited to implement DIVA strategies.

3.3. Replication-competent vs. replication-defective adenoviral vectors

When Ad are engineered to be RD and express a transgene, most of the immune response they trigger can be biased toward this transgene since transgene expression replaces early adenoviral gene expression, thus limiting adenoviral protein synthesis [24]. Ad can also be engineered to express transgene while remaining replication competent (RC). In these cases, immune responses to the transgene can be enhanced [9, 31, 40], but the immune system is also more prone to react to the vector than in the case of RD vectors since infective lytic cycles occur. This can result in sero-neutralization of the vector over time that limits vaccine efficacy if booster immunizations are required. Care should also be taken when immunizing immunocompromised individuals with RCAd vectors as vaccine-derived pathology could be induced. Importantly, RCAd could potentially escape the vaccinated host, which limits their application and hinders their approval by legislative bodies. RCAd can nonetheless have applications in veterinary science as demonstrated by the effective campaigns for rabies control in Canada with RC adenoviral vectors expressing the rabies virus glycoprotein delivered to wildlife through baiting [41]. The vaccine was safe in a number of species and showed minimal risk of horizontal transmission [42].

The present chapter will mainly focus on RD adenoviral vectors as veterinary tools since RDAd genetic stability makes them particularly suited for the design of safe and legislatively acceptable vaccines. Despite being one of the most studied recombinant vectors in veterinary medicine, no RDAd vaccine is currently licensed for veterinary use. An RDHuAd5 vector expressing the FMDV P1 region and the $3C^{pro}$ protease has nonetheless received a conditional US veterinary biological product license. An exhaustive safety study for the issue of a US veterinary biological license product for this vaccine was recently completed [43]. No evidence of reversion to virulence, shedding from vaccinees or presence in milk products was detected indicating that RDAd vaccines are safe and recombinant vaccine particles are unlikely to be found in animal products used for human consumption.

3.4. Human vs. nonhuman adenovirus for veterinary use

HuAd5 vector is the most extensively used adenoviral vector for vaccine design and gene therapy. However, pre-existing adenoviral immunity complicates its use in human therapy since this drastically decreases efficacy [44], but in veterinary medicine, no immunity to HuAd should be present. Indeed pre-existing neutralizing antibodies and cell-mediated immunity to the veterinary specie Ad usually do not cross-react with human adenoviral vectors [45]. This implicates that human adenoviral vectors can trigger strong immune response in the veterinary host. There are nonetheless risks that need assessment prior to commercial release like reversion to virulence. Importantly for livestock animals, it is essential to demonstrate that the recombinant vaccine is absent from the animal products consumed by the human population (e.g., meat and milk) so that veterinary use of RDHuAd vaccines is not perceived as a health risk by legislative bodies and the public in general.

To circumvent pre-existing immunity, nonhuman adenoviral vectors can be used. These are often studied for gene therapy as they improve gene delivery and expression [46], but they could still hold veterinary vaccine applications. For instance, in the cases of zoonosis like Rift Valley fever (RVF) that affect human populations, it could prove advantageous to develop adenoviral-based vaccines on the backbone of nonhuman species to avoid HuAd pre-existing immunity [47]. Since most nonprimate adenoviral vectors produce abortive infections in human cells [48], the risk of virulence reversion and recombinant vector spreading in humans is further minimized. These nonhuman vectors also produce strong immune responses in the veterinary host, although most studies thus far have used RCAd constructs [9, 49, 50]. Nonhuman RCAd could have applications in veterinary vaccination when the Ad itself is pathogenic [51]. Recombinant technology could be used to attenuate pathogenic fowl adenoviruses (FoAd) strains to produce suitable vaccine strains or FoAd could be manipulated to become vectors that express recombinant immunogenic proteins [52]. Because of the dissemination risks posed by RCAd, RDAd appear nonetheless as the way forward even for nonhuman Ad.

One of the main barriers for the development of nonhuman RDAd vectors is the necessity to construct cell lines capable of complementing the viral genome so that these vaccines can be propagated. The production of RD vectors has nonetheless been achieved for several nonprimate species [48, 53], and RDCaAd2 vectors expressing immunogenic viral subunits have shown potential for vaccination against rabies [54], bluetongue virus (BTV) [55], or FMDV [56]. Because Ad infect a wide range of mammalian cells from different species, these nonhuman RDAd vectors could also be used to circumvent pre-existing immunity. Ultimately, this could help broaden the range of adenoviral vectors available for vaccine design. Understanding nonhuman adenovirus biology and advancing in their manipulation can, therefore, help vaccinologist design novel strategies in veterinary medicine and in human medicine where pre-existing immunity to these vectors will be minimal.

4. Applications of RDAd in veterinary medicine

Typically, RDAd are engineered to express an immunogenic antigen from the pathogen and used as vaccine. However, since RDAd can accommodate fairly large inserts, they can encode for multiple genes and produce virus-like particles. RDAd can also be used to boost adjuvancy in vaccine preparations by expressing cytokines or co-stimulatory molecules, or even impair viral replication by encoding for interfering RNA sequences.

4.1. Antigen-encoding RDAd as vaccines

RDAd encoding for immunogenic determinants showed promising vaccination results in a range of relevant veterinary diseases (**Table 1**). In PPRV, which is the next disease targeted by the World Organization for Animal Health (OIE) for eradication, RDHuAd5 vectors expressing PPRV fusion protein (F) or hemagglutinin (H) induced strong cellular and humoral immunity and protected goats and sheep against virulent challenge [38, 39]. In BTV, immunizations with RDHuAd5 expressing the VP2 and/or VP7 proteins are protected from homologous challenge [57]. RDHuAd5 expressing the FMDV P1 region and the $3C^{pro}$ protease can protect swine and cattle from the disease [58]. RDAd vaccines can protect multiple mammalian hosts (sheep, goats, and cattle) from Rift Valley fever virus (RVFV) challenge, and induce immunity in camels [47]. RDAd vaccines can also protect across animal classes as an RDHuAd5 vector vaccine expressing the influenza A virus (IAV) H protected chicken from viral challenge [59]. This broad spectrum of hosts makes RDAd vaccines particularly attractive for vaccine design against zoonotic diseases.

The choice of antigen is of prime importance for RDAd vaccine clinical efficiency. The immunogenicity of the transgene influences the immunity triggered to the vector [24, 31]. Strongly, immunogenic transgene products skew the immune response toward these proteins, whereas weakly immunogenic transgene products favor anti-vector immunity that eliminates transduced cells and shortens antigen exposure [60]. For instance, RDAd vaccine expressing only the

Adenovirus*	Disease	Transgene	Model/natural host	Efficacy/findings	References
Antigen encoding					
HuAd5	IAV	HA	Swine	Protection in homol-challenge	[94]
				Partial in heterol-challenge	
HuAd5	IAV	HA	Mouse poultry	Protection	[59]
				Ab + CMI	
				s.c. vaccinated chicken protected	
ChAdY25	RVFV	Gn, Gc	Sheep	Multispecies protection	[47]
			Cattle	VNA induction in camels	
			Goats		
			Camels		
HuAd5	PPRV	F, H	Sheep	Protection	[38]
				VNA, Ab production, CMI	
HuAd5	BTV	VP2, VP7	Mouse	VNA, Ab production CMI and protection	[57]
			Sheep		
HuAd5	FMDV	poGMCSF, VP1, VP1 epitopes	Mouse	Protection	[95]
			Guinea pigs		
			Swine		
HuAd5	CSFV	E2 protein	Swine	Complete protection in DNA-Ad prime boost	[67]
VLP encoding					
RC CaAd2	RHDV	VP60	Rabbit	Protection Ab production	[50]
CaAd2	FMDV	P1/3C	Guinea pigs	Ab production and protection	[56]
HuAd5	FMDV	PPV-VP2 expressing FMDV VP1 epitopes	Mouse	Protection	[96]
			Swine	VNA production	
HuAd5	FMDV	P1/3Cpro	Swine	Protection	[58, 62, 63]
			Cattle		

http://dx.doi.org/10.5772/intechopen.79132

Adenovirus*	Disease	Transgene	Model/natural host	Efficacy/findings	References
RNA interference					
HuAd5	FMDV	poIFN-α + poIFN-γ + siRNA against NS proteins	Mouse	Protection	[79]
			Guinea pigs		
			Swine		
HuAd5	FMDV	shRNA	Guinea pigs	Partial protection	[78]
			Swine		
Immunomodulation					
HuAd5	FMDV	poIFN-α + poIFN-γ	Swine	Synergistic protection	[73]
HuAd5	FMDV	poIFN-α	Swine	Protection vs. several FMDV serotypes	[72]
HuAd5	IAV	ovIFN-τ	Mouse	Protection	[74]
HuAd5	Salmonella	poGCSF	Swine	Protection against Salmonella shedding and colonization	[76]
HuAd5	PCV2	poGM-CSF	Swine	Reduced viremia	[81]
		poCD40L PCV Capsid protein			
HuAd5	PRRSV	Gp3 GP5 fusion protein poCD40L	Swine	Partial protection	[97]
				Ab and CMI	
				CD40L improve efficacy	
HuAd5	PRRSV	GP3 GP5 fusion protein, HSP70	Swine	IFN-γ and IL-4 in sera VNA HSP70 improves efficacy	[83]

*All adenoviral vectors are replication deficient unless otherwise stated (i.e., RC).

Table 1. Examples of adenoviral vector use in veterinary medicine.

FMDV VP1 capsid protein can only induce low levels of neutralizing antibodies [61], whereas RDAd vaccines expressing the complete P1-encoded capsid polypeptide of FMDV and the $3C^{pro}$ protease can fully protect swine and cattle [58, 62, 63]. Protection was also achieved in animal models with this FMDV antigen formulation expressed in an RDCaAd2 vector instead of the "traditional" RDHuAd5 vector [56], highlighting the efficacy of this antigen construct. The choice of antigen for vaccination should, therefore, be based on the knowledge of host-pathogen interactions and the characterization of the protective immunity that arises during infection.

Typically, RDAd are very effective at triggering cell-mediated immunity, since transduction allows for prolonged presentation of intracellular antigen encoded by the transgene. This can be very useful for vaccine design, and inclusion of genes targeted by cell-mediated immunity could improve immunogenicity [57, 64]. Cell-mediated immunity can target epitopes encoded by conserved genes and thereby recognize infected cells independently of the virus serotype [65]. This could potentially provide some degree of protection against heterologous serotypes [66] in diseases like FMDV, IAV, or BTV in which cross-protection between serotypes is very limited. Inclusion of immunogenic antigens for cell immunity will likely improve RDAd vaccine efficacy.

RDAd vector expressing antigens are nonetheless fully protective in only few cases. Ideally, a veterinary vaccine should consist of a single-dose immunization that provides long-term protection so that costs are maintained low. Some RDAd vaccines can achieve this [58, 66], but experimental vaccination protocols often employ prime-boost strategies for RDAd vaccines to trigger protective immunity. In some cases, prime-boost strategies appear necessary to RDAd vaccine activity [67]. Administration route can also affect RDAd vaccine efficacy [68], and induction of mucosal immunity can be limited. Oral/nasal RDAd administration can nonetheless trigger the mucosal immunity necessary for protection against influenza for instance [66, 69]. RDAd administration route should, therefore, be given careful attention when designing vaccination protocol.

4.2. Immunomodulation through RDAd vectors

Enhancing the immunogenicity of RDAd vaccine candidates so that efficacy is improved is a continuous goal for researchers. This could be achieved through addition of external adjuvant [70], or by making the adenoviral vector encode for immunomodulatory molecules that would favor immune response to the antigen (**Table 1**).

The antiviral activity of the IFN system is well documented [71]. IFNs induce an antiviral state in cells that help the host control viral infections. Systemic administration of recombinant IFNs is nonetheless toxic and too expensive for veterinary medicine. As an alternative, inclusion of IFNs as RDAd transgenes could boost vaccine efficacy and/or provide early protection when highly contagious virus outbreaks occur. Recombinant expression of IFN-α with FMDV VP1 protein or epitopes enhanced the RDAd vaccine activity [61]. IFN-expressing RDAd have nonetheless shown their potential as antiviral agents when administered on their own. RDAd expressing porcine IFN-α can protect against multiple FMDV serotypes [72] and work synergistically with IFN-γ to protect against FMDV challenge [73]. Ovine IFN-τ expression in RDAd demonstrated antiviral efficacy in influenza virus murine model [74]. This ruminant IFN displays many of the antiviral activities of IFN-α in a wide range of mammalian hosts but with reduced toxicity [75]. IFN-expressing RDAd have, therefore, the potential to be used as off-the-shelf antiviral agents in the early stages of an outbreak in a disease-free country that could control disease spread for highly contagious viral pathogens like FMDV. They can also help bridge the gap in immunity in naïve herds, while the adaptive immune response to the vaccine is being triggered. Cytokine expression by RDAd could also have applications for the treatment of bacterial infections, since for instance, RDAd-expressed porcine G-CSF was successful at reducing Salmonella shedding and colonization in challenged pigs [76].

4.3. RNA interference of viral replication and enhanced antigen presentation

RNA interference can be an effective mean to impair viral replication [77], and its delivery through an RDAd vector could be attractive to treat some viral diseases. Expression of small hairpin RNAs specific for the FMDV 3D polymerase and the structural 1D protein could partially protect pigs against challenge [78]. RDAd delivering small interfering RNA, IFN-α, and IFN-γ enhanced anti-FMDV effects and was effective against multiple FMDV serotypes [79]. RNA interference delivered by RDAd could, therefore, be used as a fast-acting antiviral. This strategy could complement the efficacy of IFN-expressing RDAd, as these antiviral effects act through different pathways.

Antigen expression on RDAd can be engineered to promote antigen presentation. This has been achieved for instance by linking the antigen to the invariant chain to promote antigen presentation and thus enhances cell-mediated immunity [80]. Inclusion of GM-CSF or CD40L expression in the RDAd vectors probably favors antigen presentation and improves vaccine effectiveness [81]. Antigen delivery can also be improved by expressing the antigen of interest linked to heat shock proteins. Expression of the HSP70 C-terminal gene linked to the hantavirus glycoprotein Gn can augment cellular and humoral immunity and protects mice from a virulent challenge [82]. Co-expression of HSP70 and PRRSV gp3 and gp5 glycoproteins in an RDAd vector also enhances immunity to the antigens and improves vaccine efficacy [83]. Strategies that boost transgene antigen presentation can, therefore, become a valuable tool to improve RDAd vaccine immunogenicity.

5. Safeties and risks of adenoviral vectors

Different issues such as the oncogenic or mutagenic risk of the modified vector, its origin, its tropism, or its pathogenicity are some of the potential concerns around adenoviral vector use, not only for the host but also for the environment [84]. Adenoviral vectors are classified as Risk Group 2 (RG2) agents, defined as pathogens causing infrequent serious human diseases with available prevention therapies. This group of agents has to be manipulated in a biosafety level 2 containment facility (BSL2) [85]. Gloves, eye, nose, and mouth protection and laboratory coat are required to prevent mucous membrane contact, inhalation of aerosolized droplets, ingestion, or parenteral inoculation.

Ad cause usually mild illnesses, except in immunocompromised individuals. Potential toxicity is documented *in vitro* and in *in vivo* mouse models for the first-generation RDAds, which contain a great proportion of the Ad genome [86]. These vectors also have the risk of reversion to replication competence because of recombination or complementation between the left terminus end of the vector and the partially overlapping E1 sequence present in HEK293 cell genome or in adenoviral sequences previously acquired by the host (due to the general distribution of the Ad) [87]. Packaging cell lines with nonhomologous sequences with the vector or testing viral vector stocks for RC virus can be employed to reduce this risk [88]. Deletion of the E2A, E2B or E4 regions in the second-generation vectors reduces this risk but complicates the packaging of the recombinant adenoviral particle since specific packaging cells have to be designed to complement the missing adenoviral genome. Obtaining high titer stocks with these systems is more difficult, which often leads to reduced immunogenicity as only lower vaccine doses can be obtained [89]. These issues are even more pronounced with "gutless" adenoviral vectors, which are perfect in terms of safety, but can be problematic in terms of immunogenicity and ease of production.

The route of administration is also relevant in RDAd shedding. Intravenous (or systemic) administration results predominantly in liver adenovirus localization with minimal or no shedding to biologic fluids [6, 90]. When administered subcutaneously or intramuscularly, the point of inoculation should be disinfected to minimize the risk of vector propagation to the environment as leaks can sometimes be detected at the site. Nonetheless, no vector was detected in rodents 72 h after injection in tail swab, and the vectors were cleared from blood within 24 h [90]. As previously mentioned, RDAd vectors do not integrate efficiently into the host cell genome, the transgene expression is only transient and they do not produce infective particles, which inherently improves their biosafety [8]. Adenoviral vector survival in bedding or caging is also reduced compared to parent Ad [8]. Adenoviral vectors can, however, trigger episodes of inflammatory responses. This includes one death after a high dose direct injection of an adenoviral vector into the hepatic artery [91], which produced a fulminant immune reaction probably due to pre-existing vector immunity. It is, however, very difficult to detect vertical or germline transmission of adenovirus vectors in experimental animal models [92].

All measures (autoclave treatments for 30 min at 121°C under 1 atm pressure, 0.5% sodium hypochlorite, 5% phenol, or 2% glutaraldehyde) sufficient to eliminate the peril of adenoviral transmission have to be met to minimize risks (alcohol is not a good decontaminant for Ad), but we must not forget that RDAd do not replicate and should not, unless recombination and complementation occur, be able to shed from inoculated animals, and are thus even less likely to infect another organism. In the case of an RCAd, the risk is reduced to the range and tropism of the Ad; for example, human adenovirus is only known to replicate in two nonhuman species: cotton rat and hamster [93].

It is necessary to deepen in the knowledge of the biodistribution, dissemination, and *in vivo* transgene expression duration of these vectors in veterinary medicine to assess their risk more thoroughly. No standard procedures to monitor these risks exist, and thus, each independent study analyses arbitrarily which biosafety parameters are evaluated.

6. Conclusions and perspectives

In the increasingly globalized world in which we live, animal health is of great importance and the prevention of animal diseases through vaccination is necessary for animal care, food production, food safety, food security, prevention of zoonotic and foodborne infections, reduction of antibiotic needs, and public health. That vaccination is an integral part of global disease prevention, which can even eradicate diseases is a fact. We have examples of this in both human and animal health with the eradication of smallpox and rinderpest. However, there are still many animal diseases without vaccines or for which treatment needs improvement. Numerous studies constructing, testing, characterizing, optimizing, and identifying adenoviral-based vaccines as optimal against different animal diseases appeared in the last decades. They elicit potent cellular and humoral immunity and can be implemented along DIVA diagnostic tests. RDAd can also be used to deliver immunomodulation to improve disease treatment. Transference to the veterinary market is, however, lagging behind laboratory advances, and no adenoviral vector-based vaccine has yet obtained a veterinary license for systematic use in the field. This nonetheless appears nowadays closer with the recent publication of a positive safety report on an RDHuAd5 FMDV vaccine [43]. Recombinant RDAd reagents could, therefore, have great economic relevance in the future in veterinary medicine. Regulatory committees both in the EU and in the US should favor the approval of these reagents, based on the increasing scientific evidence for their efficacy and safety so that recombinant RDAd can make the leap from laboratory to the field. At the moment, the regulatory bases (EMEA/CVMP/004/04) for the use of adenoviral vector-based vaccines in farms are not well defined, although there are bases established in the EU by the European Medicine Agency (EMEA) and its Committee for Veterinary medicinal Products (CVMP) and in the US by the Animal and Plant Health Inspection Service (APHIS) from The United States Department of Agriculture (USDA). A global cooperation between the veterinary industry and governments is needed in the future for adenoviral vector-based vaccines to reach the market.

Acknowledgements

The work in the lab was funded by Grants RyC2010-06516, AGL2011-25025, AGL2012-33289, AGL2015-64290R, ADENONET-Redes de Excelencia (BIO2015-68990-REDT) from the Spanish Ministerio de Economía y Competitividad, and Grant S2013/ABI-2906-PLATESA from the Comunidad de Madrid and the European Union (Fondo Europeo de Desarrollo Regional, FEDER funds) and 731014-VetBioNet Project from European Union.

Conflict of interest

The authors declare no conflict of interest.

Abbreviations

Ab	antibody
Ad	adenovirus
APHIS	animal and plant health inspection service
atm	atmosphere

BTV	bluetongue virus
Bo	bovine
BSL2	biosafety level 2 containment facility
°C	centigrade
Ca	canine
CDV	canine distemper virus
CMI	cell-mediated immunity
Ch	chimpanzee
CSFV	classical swine fever
CVMP	committee for veterinary medical products
DC	dendritic cells
DNA	deoxyribonucleic acid
DIVA	differentiating infected from vaccinated animals
EMA	European medical agency
EU	European Union
FMDV	foot and mouth disease
Fo	fowl
F	fusion protein
Gn, Gc	glycoproteins
GM-CSF	granulocyte macrophage colony-stimulating factor
HEK	human embryonic kidney
H-HA	hemagglutinin
HV	herpes virus
HSP	heat shock protein
HuAd	human adenovirus
IAV	influenza A virus
IFN	interferon
IL	interleukin
IPS-1	interferon-beta promoter stimulator-1
LRR	leucine-rich-repeat
MIP	macrophage inflammatory protein
MyD88	myeloid differentiation protein 88

N	nucleoprotein
NACHT	neuronal apoptosis inhibitor protein (NAIP), class 2 transcription activator of the MHC (C2TA), heterokaryon incompatibility (HET-E) and telomerase-associated protein 1 (TP1)
NALP3	NACHT, LRR, and PYD domains-containing
NIH	National Institutes of health
NLR	nucleotide-binding oligomerization domain-like receptor
Ov	ovine
PCV	porcine circovirus
PYD	"PYRIN domain," after the *pyrin* proteins
PPR	peste des petits ruminants
PPRV	peste des petits ruminants virus
Po	porcine
PRRSV	porcine reproductive and respiratory syndrome virus
RHDV	rabbit hemorrhagic disease virus
RD	replication-defective
RC	replication-competent
RVF	Rift Valley fever
RVFV	Rift Valley fever virus
RG2	Risk Group 2
TP	terminal protein
TLR	toll-like receptor
US	United States
USDA	United States Department of Agriculture
VNA	virus neutralizing antibodies
OIE	World Organization for Animal Health

Author details

José M. Rojas, Noemí Sevilla and Verónica Martín*

*Address all correspondence to: veronica.martin@inia.es

Centro de Investigación en Sanidad Animal, Instituto Nacional de Investigación y Tecnología Agraria y Alimentaria (CISA-INIA), Valdeolmos, Madrid, Spain

References

[1] Rowe WP, Huebner RJ, Gilmore LK, Parrott RH, Ward TG. Isolation of a cytopathogenic agent from human adenoids undergoing spontaneous degeneration in tissue culture. Proceedings of the Society for Experimental Biology and Medicine. 1953;**84**:570-573

[2] Davison AJ, Benko M, Harrach B. Genetic content and evolution of adenoviruses. Journal of General Virology. 2003;**84**:2895-2908. DOI: 10.1099/vir.0.19497-0

[3] Murray PR, Rosenthal KS, Pfaller MA. Adenoviruses. 8th ed. Philadelphia, PA: Microbiología Médica; 2016. pp. 418-424. ISBN: 19103-2899

[4] Jones MS, Harrach B, Ganac RD, Gozum MMA, Cruz Dela WP, Riedel B, et al. New adenovirus species found in a patient presenting with gastroenteritis. Journal of Virology. 2007;**81**:5978-5784. DOI: 10.1128/JVI.02650-06

[5] Ross PJ, Parks RJ. Construction and Characterization of Adenovirus Vectors. NY, USA: Cold Spring Harb Protoc; 2009. DOI: 10.1101/pdb.prot5011

[6] Khare R, Chen CY, Weaver EA, Barry MA. Advances and future challenges in adenoviral vector pharmacology and targeting. Current Gene Therapy. 2011;**11**:241-258. DOI: 10.2174/156652311796150363

[7] Gonçalves MAFV, de Vries AAF. Adenovirus: From foe to friend. Reviews in Medical Virology. 2006;**16**:167-186. DOI: 10.1002/rmv.494

[8] Chuah MKL, Collen D, VandenDriessche T. Biosafety of adenoviral vectors. Current Gene Therapy. 2003;**3**:527-543

[9] Reddy PS, Idamakanti N, Pyne C, Zakhartchouk AN, Godson DL, Papp Z, et al. The immunogenicity and efficacy of replication-defective and replication-competent bovine adenovirus-3 expressing bovine herpesvirus-1 glycoprotein gD in cattle. Veterinary Immunology and Immunopathology. 2000;**76**:257-268

[10] Fischer L, Tronel JP, Pardo-David C, Tanner P, Colombet G, Minke J, et al. Vaccination of puppies born to immune dams with a canine adenovirus-based vaccine protects against a canine distemper virus challenge. Vaccine. 2002;**20**:3485-3497

[11] Reddy PS, Idamakanti N, Babiuk LA, Mehtali M, Tikoo SK. Porcine adenovirus-3 as a helper-dependent expression vector. Journal of General Virology. 1999;**80**(Pt 11):2909-2916. DOI: 10.1099/0022-1317-80-11-2909

[12] Ertl HC. Viral vectors as vaccine carriers. Current Opinion in Virology. 2016;**21**:1-8. DOI: 10.1016/j.coviro.2016.06.001

[13] Yang TC, Dayball K, Wan YH, Bramson J. Detailed analysis of the $CD8^+$ T-cell response following adenovirus vaccination. Journal of Virology. 2003;**77**:13407-13411. DOI: 10.1128/JVI.77.24.13407-13411.2003

[14] Hartman ZC, Kiang A, Everett RS, Serra D, Yang XY, Clay TM, et al. Adenovirus infection triggers a rapid, MyD88-regulated transcriptome response critical to acute-phase and adaptive immune responses in vivo. Journal of Virology. 2007;**81**:1796-1812. DOI: 10.1128/JVI.01936-06

[15] Rhee EG, Blattman JN, Kasturi SP, Kelley RP, Kaufman DR, Lynch DM, et al. Multiple innate immune pathways contribute to the immunogenicity of recombinant adenovirus vaccine vectors. Journal of Virology. 2011;**85**:315-323. DOI: 10.1128/JVI.01597-10

[16] Perreau M, Welles HC, Pellaton C, Gjoksi B, Potin L, Martin R, et al. The number of Toll-like receptor 9-agonist motifs in the adenovirus genome correlates with induction of dendritic cell maturation by adenovirus immune complexes. Journal of Virology. 2012;**86**:6279-6285. DOI: 10.1128/JVI.00123-12

[17] Hemmi M, Tachibana M, Tsuzuki S, Shoji M, Sakurai F, Kawabata K, et al. The early activation of $CD8^+$ T cells is dependent on type I IFN signaling following intramuscular vaccination of adenovirus vector. BioMed Research International. 2014;**2014**:158128. DOI: 10.1155/2014/158128

[18] Broz P, Dixit VM. Inflammasomes: Mechanism of assembly, regulation and signalling. Nature Reviews Immunology. 2016;**16**:407-420. DOI: 10.1038/nri.2016.58

[19] Muruve DA, Pétrilli V, Zaiss AK, White LR, Clark SA, Ross PJ, et al. The inflammasome recognizes cytosolic microbial and host DNA and triggers an innate immune response. Nature. 2008;**452**:103-107. DOI: 10.1038/nature06664

[20] Lieber A, He CY, Meuse L, Schowalter D, Kirillova I, Winther B, et al. The role of Kupffer cell activation and viral gene expression in early liver toxicity after infusion of recombinant adenovirus vectors. Journal of Virology. 1997;**71**:8798-8807

[21] Di Paolo NC, Miao EA, Iwakura Y, Murali-Krishna K, Aderem A, Flavell RA, et al. Virus binding to a plasma membrane receptor triggers interleukin-1 alpha-mediated proinflammatory macrophage response in vivo. Immunity. 2009;**31**:110-121. DOI: 10.1016/j.immuni.2009.04.015

[22] Lindsay RWB, Darrah PA, Quinn KM, Wille-Reece U, Mattei LM, Iwasaki A, et al. $CD8^+$ T cell responses following replication-defective adenovirus serotype 5 immunization are dependent on $CD11c^+$ dendritic cells but show redundancy in their requirement of TLR and nucleotide-binding oligomerization domain-like receptor signaling. Journal of Immunology. 2010;**185**:1513-1521. DOI: 10.4049/jimmunol.1000338

[23] Quinn KM, Zak DE, Costa A, Yamamoto A, Kastenmuller K, Hill BJ, et al. Antigen expression determines adenoviral vaccine potency independent of IFN and STING signaling. Journal of Clinical Investigation. 2015;**125**:1129-1146. DOI: 10.1172/JCI78280

[24] Schagen FHE, Ossevoort M, Toes REM, Hoeben RC. Immune responses against adenoviral vectors and their transgene products: A review of strategies for evasion. Critical Reviews in Oncology/Hematology. 2004;**50**:51-70. DOI: 10.1016/S1040-8428(03)00172-0

[25] Heemskerk B, Veltrop-Duits LA, van Vreeswijk T, Dam ten MM, Heidt S, Toes REM, et al. Extensive cross-reactivity of $CD4^+$ adenovirus-specific T cells: Implications for immunotherapy and gene therapy. Journal of Virology. 2003;**77**:6562-6566. DOI: 10.1128/JVI.77.11.6562-6566.2003

[26] Xiang Z, Gao G, Reyes-Sandoval A, Cohen CJ, Li Y, Bergelson JM, et al. Novel, chimpanzee serotype 68-based adenoviral vaccine carrier for induction of antibodies to a transgene product. Journal of Virology. 2002;**76**:2667-2675. DOI: 10.1128/JVI.76.6.2667-2675.2002

[27] Small JC, Haut LH, Bian A, Ertl HCJ. The effect of adenovirus-specific antibodies on adenoviral vector-induced, transgene product-specific T cell responses. Journal of Leukocyte Biology. 2014;**96**:821-831. DOI: 10.1189/jlb.1A0813-451RR

[28] Baron MD, Iqbal M, Nair V. Recent advances in viral vectors in veterinary vaccinology. Current Opinion in Virology. 2018;**29**:1-7. DOI: 10.1016/j.coviro.2018.02.002

[29] Shiver JW, Fu T-M, Chen L, Casimiro DR, Davies M-E, Evans RK, et al. Replication-incompetent adenoviral vaccine vector elicits effective anti-immunodeficiency-virus immunity. Nature. 2002;**415**:331-335. DOI: 10.1038/415331a

[30] Ferreira TB, Alves PM, Aunins JG, Carrondo MJT. Use of adenoviral vectors as veterinary vaccines. Gene Therapy. 2005;**12**(Suppl 1):S73-S83. DOI: 10.1038/sj.gt.3302618

[31] Fougeroux C, Holst PJ. Future prospects for the development of cost-effective adenovirus vaccines. International Journal of Molecular Sciences. 2017;**18**(4):686. DOI: 10.3390/ijms18030686

[32] Hartman ZC, Appledorn DM, Amalfitano A. Adenovirus vector induced innate immune responses: impact upon efficacy and toxicity in gene therapy and vaccine applications. Virus Research. 2008;**132**:1-14. DOI: 10.1016/j.virusres.2007.10.005

[33] Tandon M, Sharma A, Vemula SV, Bangari DS, Mittal SK. Sequential administration of bovine and human adenovirus vectors to overcome vector immunity in an immunocompetent mouse model of breast cancer. Virus Research. 2012;**163**:202-211. DOI: 10.1016/j.virusres.2011.09.031

[34] Kallel H, Kamen AA. Large-scale adenovirus and poxvirus-vectored vaccine manufacturing to enable clinical trials. Biotechnology Journal. 2015;**10**:741-747. DOI: 10.1002/biot.201400390

[35] Pelliccia M, Andreozzi P, Paulose J, D'Alicarnasso M, Cagno V, Donalisio M, et al. Additives for vaccine storage to improve thermal stability of adenoviruses from hours to months. Nature Communications. 2016;**7**:13520. DOI: 10.1038/ncomms13520

[36] Diaz-San Segundo F, Medina GN, Stenfeldt C, Arzt J, de los Santos T. Foot-and-mouth disease vaccines. Veterinary Microbiology. 2017;**206**:102-112. DOI: 10.1016/j.vetmic.2016.12.018

[37] Rojas JM, Moreno H, García A, Ramírez JC, Sevilla N, Martín V. Two replication-defective adenoviral vaccine vectors for the induction of immune responses to PPRV. Vaccine. 2014;**32**:393-400. DOI: 10.1016/j.vaccine.2013.11.033

[38] Rojas JM, Moreno H, Valcárcel F, Peña L, Sevilla N, Martín V. Vaccination with recombinant adenoviruses expressing the peste des petits ruminants virus F or H proteins overcomes viral immunosuppression and induces protective immunity against PPRV challenge in sheep. PLoS One. 2014;**9**:e101226. DOI: 10.1371/journal.pone.0101226

[39] Herbert R, Baron J, Batten C, Baron M, Taylor G. Recombinant adenovirus expressing the haemagglutinin of Peste des petits ruminants virus (PPRV) protects goats against challenge with pathogenic virus; a DIVA vaccine for PPR. Veterinary Research. 2014;**45**:24. DOI: 10.1186/1297-9716-45-24

[40] Peng B, Wang LR, Gómez-Román VR, Davis-Warren A, Montefiori DC, Kalyanaraman VS, et al. Replicating rather than nonreplicating adenovirus-human immunodeficiency virus recombinant vaccines are better at eliciting potent cellular immunity and priming high-titer antibodies. Journal of Virology. 2005;**79**:10200-10209. DOI: 10.1128/JVI.79.16.10200-10209.2005

[41] Rosatte RC, Donovan D, Davies JC, Brown L, Allan M, Zuben von V, et al. High-density baiting with ONRAB® rabies vaccine baits to control Arctic-variant rabies in striped skunks in Ontario, Canada. Journal of Wildlife Diseases. 2011;**47**:459-465. DOI: 10.7589/0090-3558-47.2.459

[42] Knowles MK, Nadin-Davis SA, Sheen M, Rosatte R, Mueller R, Beresford A. Safety studies on an adenovirus recombinant vaccine for rabies (AdRG1.3-ONRAB) in target and non-target species. Vaccine. 2009;**27**:6619-6626. DOI: 10.1016/j.vaccine.2009.08.005

[43] Barrera J, Brake DA, Kamicker BJ, Purcell C, Kaptur R, Schieber T, et al. Safety profile of a replication-deficient human adenovirus-vectored foot-and-mouth disease virus serotype A24 subunit vaccine in cattle. Transboundary and Emerging Diseases. 2018;**65**:447-455. DOI: 10.1111/tbed.12724

[44] Shiver JW, Emini EA. Recent advances in the development of HIV-1 vaccines using replication-incompetent adenovirus vectors. Annual Review of Medicine. 2004;**55**:355-372. DOI: 10.1146/annurev.med.55.091902.104344

[45] Sharma A, Tandon M, Ahi YS, Bangari DS, Vemulapalli R, Mittal SK. Evaluation of cross-reactive cell-mediated immune responses among human, bovine and porcine adenoviruses. Gene Therapy. 2010;**17**:634-642. DOI: 10.1038/gt.2010.1

[46] Lopez-Gordo E, Podgorski II, Downes N, Alemany R. Circumventing antivector immunity: potential use of nonhuman adenoviral vectors. Human Gene Therapy. 2014;**25**: 285-300. DOI: 10.1089/hum.2013.228

[47] Warimwe GM, Gesharisha J, Carr BV, Otieno S, Otingah K, Wright D, et al. Chimpanzee adenovirus vaccine provides multispecies protection against Rift Valley Fever. Scientific Reports. 2016;**6**:20617. DOI: 10.1038/srep20617

[48] van Olphen AL, Tikoo SK, Mittal SK. Characterization of bovine adenovirus type 3 E1 proteins and isolation of E1-expressing cell lines. Virology. 2002;**295**:108-118. DOI: 10.1006/viro.2002.1389

[49] Qin J, Huang H, Ruan Y, Hou X, Yang S, Wang C, et al. A novel recombinant Peste des petits ruminants-canine adenovirus vaccine elicits long-lasting neutralizing antibody response against PPR in goats. PLoS One. 2012;**7**:e37170. DOI: 10.1371/journal.pone.0037170

[50] Jiang Q, Yu Z, Liu J-S, Kong D-S, Guo D-C, Quan C-S, et al. Recombinant canine adenovirus type 2 expressing rabbit hemorrhagic disease virus VP60 protein provided protection against RHD in rabbits. Veterinary Microbiology. 2018;**213**:15-20. DOI: 10.1016/j.vetmic.2017.11.007

[51] Brown Jordan A, Gongora V, Hartley D, Oura C. A review of eight high-priority, economically important viral pathogens of poultry within the Caribbean Region. Veterinary Science. 2018;**5**(2):51. DOI: 10.3390/vetsci5010014

[52] Pei Y, Corredor JC, Griffin BD, Krell PJ, Nagy É. Fowl adenovirus 4 (FAdV-4)-based infectious clone for vaccine vector development and viral gene function studies. Viruses. 2018;**10**(2):97. DOI: 10.3390/v10020097

[53] Zhang P, Du E, Ma J, Wang W, Zhang L, Tikoo SK, et al. A novel and simple method for rapid generation of recombinant porcine adenoviral vectors for transgene expression. PLoS One. 2015;**10**:e0127958. DOI: 10.1371/journal.pone.0127958

[54] Bouet-Cararo C, Contreras V, Fournier A, Jallet C, Guibert JM, Dubois E, et al. Canine adenoviruses elicit both humoral and cell-mediated immune responses against rabies following immunisation of sheep. Vaccine. 2011;**29**:1304-1310. DOI: 10.1016/j.vaccine.2010.11.068

[55] Bouet-Cararo C, Contreras V, Caruso A, Top S, Szelechowski M, Bergeron C, et al. Expression of VP7, a bluetongue virus group specific antigen by viral vectors: Analysis of the induced immune responses and evaluation of protective potential in sheep. PLoS One. 2014;**9**:e111605. DOI: 10.1371/journal.pone.0111605

[56] De Vleeschauwer AR, Zhou X, Lefebvre DJ, Garnier A, Watier F, Pignon C, et al. A canine adenovirus type 2 vaccine vector confers protection against foot-and-mouth disease in guinea pigs. Vaccine. 2018;**36**:2193-2198. DOI: 10.1016/j.vaccine.2018.02.074

[57] Martín V, Pascual E, Avia M, Peña L, Valcárcel F, Sevilla N. Protective efficacy in sheep of adenovirus-vectored vaccines against bluetongue virus is associated with specific T cell responses. PLoS One. 2015;**10**:e0143273. DOI: 10.1371/journal.pone.0143273

[58] Schutta C, Barrera J, Pisano M, Zsak L, Grubman MJ, Mayr GA, et al. Multiple efficacy studies of an adenovirus-vectored foot-and-mouth disease virus serotype A24 subunit vaccine in cattle using homologous challenge. Vaccine. 2016;**34**:3214-3220. DOI: 10.1016/j.vaccine.2015.12.018

[59] Gao W, Soloff AC, Lu X, Montecalvo A, Nguyen DC, Matsuoka Y, et al. Protection of mice and poultry from lethal H5N1 avian influenza virus through adenovirus-based immunization. Journal of Virology. 2006;**80**:1959-1964. DOI: 10.1128/JVI.80.4.1959-1964.2006

[60] Schöne D, Hrycak CP, Windmann S, Lapuente D, Dittmer U, Tenbusch M, et al. Immunodominance of adenovirus-derived $CD8^+$ T cell epitopes interferes with the induction of transgene-specific immunity in adenovirus-based immunization. Journal of Virology. 2017;**91**(20):e01184-17. DOI: 10.1128/JVI.01184-17

[61] Du Y, Dai J, Li Y, Li C, Qi J, Duan S, et al. Immune responses of recombinant adenovirus co-expressing VP1 of foot-and-mouth disease virus and porcine interferon alpha in mice and guinea pigs. Veterinary Immunology and Immunopathology. 2008;**124**:274-283. DOI: 10.1016/j.vetimm.2008.04.011

[62] Moraes MP, Mayr GA, Mason PW, Grubman MJ. Early protection against homologous challenge after a single dose of replication-defective human adenovirus type

5 expressing capsid proteins of foot-and-mouth disease virus (FMDV) strain A24. Vaccine. 2002;**20**: 1631-1639

[63] Pacheco JM, Brum MCS, Moraes MP, Golde WT, Grubman MJ. Rapid protection of cattle from direct challenge with foot-and-mouth disease virus (FMDV) by a single inoculation with an adenovirus-vectored FMDV subunit vaccine. Virology. 2005;**337**:205-209. DOI: 10.1016/j.virol.2005.04.014

[64] Moraes MP, Segundo FD-S, Dias CC, Pena L, Grubman MJ. Increased efficacy of an adenovirus-vectored foot-and-mouth disease capsid subunit vaccine expressing nonstructural protein 2B is associated with a specific T cell response. Vaccine. 2011;**29**:9431-9440. DOI: 10.1016/j.vaccine.2011.10.037

[65] Rojas JM, Peña L, Martín V, Sevilla N. Ovine and murine T cell epitopes from the nonstructural protein 1 (NS1) of bluetongue virus serotype 8 (BTV-8) are shared among viral serotypes. Veterinary Research. 2014;**45**:30. DOI: 10.1186/1297-9716-45-30

[66] Price GE, Soboleski MR, Lo C-Y, Misplon JA, Quirion MR, Houser KV, et al. Single-dose mucosal immunization with a candidate universal influenza vaccine provides rapid protection from virulent H5N1, H3N2 and H1N1 viruses. PLoS One. 2010;**5**:e13162. DOI: 10.1371/journal.pone.0013162

[67] Sun Y, Li N, Li H-Y, Li M, Qiu H-J. Enhanced immunity against classical swine fever in pigs induced by prime-boost immunization using an alphavirus replicon-vectored DNA vaccine and a recombinant adenovirus. Veterinary Immunology and Immunopathology. 2010;**137**:20-27. DOI: 10.1016/j.vetimm.2010.04.005

[68] Holst PJ, Ørskov C, Thomsen AR, Christensen JP. Quality of the transgene-specific $CD8^+$ T cell response induced by adenoviral vector immunization is critically influenced by virus dose and route of vaccination. Journal of Immunology. 2010;**184**:4431-4439. DOI: 10.4049/jimmunol.0900537

[69] Park KS, Lee J, Ahn SS, Byun Y-H, Seong BL, Baek YH, et al. Mucosal immunity induced by adenovirus-based H5N1 HPAI vaccine confers protection against a lethal H5N2 avian influenza virus challenge. Virology. 2009;**395**:182-189. DOI: 10.1016/j.virol.2009.09.018

[70] Barrera J, Schutta C, Pisano M, Grubman MJ, Brake DA, Miller T, et al. Use of ENABL® adjuvant to increase the potency of an adenovirus-vectored foot-and-mouth disease virus serotype A subunit vaccine. Vaccine. 2018;**36**:1078-1084. DOI: 10.1016/j.vaccine.2018.01.026

[71] Stetson DB, Medzhitov R. Type I interferons in host defense. Immunity. 2006;**25**:373-381. DOI: 10.1016/j.immuni.2006.08.007

[72] Dias CCA, Moraes MP, Segundo FD-S, de los Santos T, Grubman MJ. Porcine type I interferon rapidly protects swine against challenge with multiple serotypes of foot-and-mouth disease virus. Journal of Interferon & Cytokine Research. 2011;**31**:227-236. DOI: 10.1089/jir.2010.0055

[73] Moraes MP, de los Santos T, Koster M, Turecek T, Wang H, Andreyev VG, et al. Enhanced antiviral activity against foot-and-mouth disease virus by a combination of type I and II porcine interferons. Journal of Virology. 2007;**81**:7124-7135. DOI: 10.1128/JVI.02775-06

[74] Martín V, Pascual E, Avia M, Rangel G, de Molina A, Alejo A, et al. A recombinant adenovirus expressing ovine interferon tau prevents influenza virus-induced lethality in mice. Journal of Virology. 2016;**90**:3783-3788. DOI: 10.1128/JVI.03258-15

[75] Chon TW, Bixler S. Interferon-tau: Current applications and potential in antiviral therapy. Journal of Interferon & Cytokine Research. 2010;**30**:477-485. DOI: 10.1089/jir.2009.0089

[76] Bearson SMD, Bearson BL, Loving CL, Allen HK, Lee I, Madson D, et al. Prophylactic administration of vector-encoded porcine granulocyte-colony stimulating factor reduces Salmonella shedding, tonsil colonization, and microbiota alterations of the gastrointestinal tract in Salmonella-challenged swine. Frontiers in Veterinary Science. 2016;**3**:66. DOI: 10.3389/fvets.2016.00066

[77] Bennasser Y, Yeung ML, Jeang K-T. RNAi therapy for HIV infection: principles and practicalities. BioDrugs. 2007;**21**:17-22. DOI: 10.2165/00063030-200721010-00003

[78] Chen W, Liu M, Jiao Y, Yan W, Wei X, Chen J, et al. Adenovirus-mediated RNA interference against foot-and-mouth disease virus infection both in vitro and in vivo. Journal of Virology. 2006;**80**:3559-3566. DOI: 10.1128/JVI.80.7.3559-3566.2006

[79] Kim S-M, Park J-H, Lee K-N, Kim S-K, You S-H, Kim T, et al. Robust Protection against highly virulent foot-and-mouth disease virus in swine by combination treatment with recombinant adenoviruses expressing porcine alpha and gamma interferons and multiple small interfering RNAs. Journal of Virology. 2015;**89**:8267-8279. DOI: 10.1128/JVI.00766-15

[80] Spencer AJ, Cottingham MG, Jenks JA, Longley RJ, Capone S, Colloca S, et al. Enhanced vaccine-induced $CD8^+$ T cell responses to malaria antigen ME-TRAP by fusion to MHC class ii invariant chain. PLoS One. 2014;**9**:e100538. DOI: 10.1371/journal.pone.0100538

[81] Li D, Huang Y, Du Q, Wang Z, Chang L, Zhao X, et al. CD40 ligand and GMCSF coexpression enhance the immune responses and protective efficacy of PCV2 adenovirus vaccine. Viral Immunology. 2016;**29**:148-158. DOI: 10.1089/vim.2015.0109

[82] Cheng L, Yu L, Wu X, Li K, Wang F, Zhang L, et al. Induction of specific humoral and cellular immune responses in a mouse model following gene fusion of HSP70C and Hantaan virus Gn and S0.7 in an adenoviral vector. PLoS One. 2014;**9**:e88183. DOI: 10.1371/journal.pone.0088183

[83] Li J, Jiang P, Li Y, Wang X, Cao J, Wang X, et al. HSP70 fused with GP3 and GP5 of porcine reproductive and respiratory syndrome virus enhanced the immune responses and protective efficacy against virulent PRRSV challenge in pigs. Vaccine. 2009;**27**:825-832. DOI: 10.1016/j.vaccine.2008.11.088

[84] Collins DE, Reuter JD, Rush HG, Villano JS. Viral vector biosafety in laboratory animal research. Comparative Medicine. 2017;**67**:215-221

[85] NIH Office of Science Policy. NIH Guidelines for Research Involving Recombinant or Synthetic Nucleic Acid Molecules. April 2016:1-151. Department of Health and Human Services. http://osp.od.nih.gov

[86] Zou A, Atencio I, Huang W-M, Horn M, Ramachandra M. Overexpression of adenovirus E3-11.6K protein induces cell killing by both caspase-dependent and caspase-independent mechanisms. Virology. 2004;**326**:240-249. DOI: 10.1016/j.virol.2004.06.007

[87] Murakami P, Pungor E, Files J, Do L, van Rijnsoever R, Vogels R, et al. A single short stretch of homology between adenoviral vector and packaging cell line can give rise to cytopathic effect-inducing, helper-dependent E1-positive particles. Human Gene Therapy. 2002;**13**: 909-920. DOI: 10.1089/10430340252939023

[88] Fallaux FJ, Bout A, van der Velde I, van den Wollenberg DJ, Hehir KM, Keegan J, et al. New helper cells and matched early region 1-deleted adenovirus vectors prevent generation of replication-competent adenoviruses. Human Gene Therapy. 1998;**9**:1909-1917. DOI: 10.1089/hum.1998.9.13-1909

[89] Volpers C, Kochanek S. Adenoviral vectors for gene transfer and therapy. Journal of Gene Medicine. 2004;**6**(Suppl 1):S164-S171. DOI: 10.1002/jgm.496

[90] Reuter JD, Fang X, Ly CS, Suter KK, Gibbs D. Assessment of hazard risk associated with the intravenous use of viral vectors in rodents. Comparative Medicine. 2012;**62**:361-370

[91] Marshall E. Gene therapy death prompts review of adenovirus vector. Science. 1999;**286**: 2244-2245

[92] Gordon JW. Direct exposure of mouse ovaries and oocytes to high doses of an adenovirus gene therapy vector fails to lead to germ cell transduction. Molecular Therapy. 2001;**3**:557-564. DOI: 10.1006/mthe.2001.0290

[93] Baldo A, van den Akker E, Bergmans HE, Lim F, Pauwels K. General considerations on the biosafety of virus-derived vectors used in gene therapy and vaccination. Current Gene Therapy. 2013;**13**:385-394. DOI: 10.2174/15665232113136660005

[94] Braucher DR, Henningson JN, Loving CL, Vincent AL, Kim E, Steitz J, et al. Intranasal vaccination with replication-defective adenovirus type 5 encoding influenza virus hemagglutinin elicits protective immunity to homologous challenge and partial protection to heterologous challenge in pigs. Clinical and Vaccine Immunology. 2012;**19**:1722-1729. DOI: 10.1128/CVI.00315-12

[95] Du Y, Jiang P, Li Y, He H, Jiang W, Wang X, et al. Immune responses of two recombinant adenoviruses expressing VP1 antigens of FMDV fused with porcine granulocyte macrophage colony-stimulating factor. Vaccine. 2007;**25**:8209-8219. DOI: 10.1016/j.vaccine.2007.09.062

[96] Pan Q, Wang H, Ouyang W, Wang X, Bi Z, Xia X, et al. Immunogenicity of adenovirus-derived porcine parvovirus-like particles displaying B and T cell epitopes of foot-and-mouth disease. Vaccine. 2016;**34**:578-585. DOI: 10.1016/j.vaccine.2015.11.003

[97] Cao J, Wang X, Du Y, Li Y, Wang X, Jiang P. CD40 ligand expressed in adenovirus can improve the immunogenicity of the GP3 and GP5 of porcine reproductive and respiratory syndrome virus in swine. Vaccine. 2010;**28**:7514-7422. DOI: 10.1016/j.vaccine.2010.09.002

Chapter 4

Adenoviral Vector-Based Vaccines and Gene Therapies: Current Status and Future Prospects

Shakti Singh, Rakesh Kumar and Babita Agrawal

Additional information is available at the end of the chapter

http://dx.doi.org/10.5772/intechopen.79697

Abstract

Adenoviruses are one of the most genetically diverse DNA viruses and cause non-life-threatening infections in the ocular, respiratory, or gastrointestinal epithelium of a diverse range of hosts. Adenoviruses are excellent vectors for delivering genes or vaccine antigens to the target host tissues and are being tested in several vaccine and gene therapy studies. Adenovirus-based vectors offer several advantages over other viral vectors such as broad range of tissue tropism, well-characterized genome, ease of genetic manipulation including acceptance of large transgene DNA insertions, inherent adjuvant properties, ability to induce robust transgene-specific T cell and antibody responses, non-replicative nature in host, and ease of production at large scale. However, several studies have highlighted major drawbacks to using adenovirus as vaccine and gene therapy vectors. These include pre-existing immunity in humans, inflammatory responses, sequestering of the vector to liver and spleen, and immunodominance of the vector genes over transgenes. In the same vein, recently discovered protein sequence homology and heterologous immunity between adenoviruses and hepatitis C virus have significant implications in the use of adenoviral vectors for vaccine development, especially for hepatitis C virus. This chapter focuses on the current scope and challenges in using adenoviral vector-based vaccines and gene therapies.

Keywords: adenoviruses, DNA viruses, viral vector, vaccine, gene therapy, immunity

1. Introduction

Adenoviruses (Ads) are non-enveloped, icosahedral DNA viruses with virion size ranges between 70 and 90 nm [1]. They belong to a diverse family (>50 serotypes) of DNA viruses called adenoviridae. Adenovirus was first isolated from human adenoid tissues in 1953 by

Rowe and his colleagues [2]. Adenoviruses usually cause non-symptomatic respiratory tract infections in both human and animals but can be life-threatening to immunocompromised individuals. Certain human adenovirus serotypes are ubiquitous in children, and most adults carry neutralizing antibodies to adenoviruses [3]. Nonetheless, since their initial use in gene therapy, they have gained wide recognition as a vaccine antigen delivery vehicle and have proven to be safe and efficient vaccine vectors for eliciting protective immune responses against transgene antigens in many animal and human studies. Recently, adenovirus vectors have been employed to attack cancer cells in cancer therapy [4]. In this chapter, we introduce different adenoviruses and their biology and potential for use in gene delivery, vaccine, and therapeutics in several human diseases. In addition, we will discuss their limitations and future prospects.

2. Adenoviruses

2.1. Genome and proteins

Adenoviruses contain a 26–45 kb size double-stranded DNA genome, inside their icosahedral virion [1]. The DNA genome of adenoviruses contains two inverted terminal repeats with 100–140 bp flanks on both the ends. Due to its small genome size, adenoviruses employ several strategies to maximally utilize its genome. For example, they encode proteins from both DNA strands, employ alternate-splicing, and use different poly A modifications of its mRNA. Adenoviral genes can be divided into five early and five late genes. Once internalized into target cells, the adenoviruses express the early genes E1A, E1B, E2, E3, and E4, which modulate host gene expression required for adenovirus protein synthesis and replication. The late transcriptional units include L1–L5 and are required in the assembly, release, and lysis of host cells [1, 5, 6] (**Figure 1**).

Structurally, adenovirus consists of a core of capsid and genome. The viral capsid consists of structural proteins hexon, penton, fiber, IIIa, VIII, and IX. Hexons are major surface structural proteins consisting of 270 trimers, which are arranged as 12 pentamers of pentons at the top of 12 icosahedral vertices. Hexons also contain several hypervariable regions and are the main targets of neutralizing antibodies. In adenoviral vectors, these sites can be engineered to carry vaccine antigen. Each icosahedral vertex gives rise to protruding fibers consisting of 12 trimers. Both penton and fiber proteins serve as ligands for host cell receptors and help in viral entry. The IIIa proteins are located in the inner surface of the capsid and help in the assembly and stabilization of vertex regions and also in the assembly of packaged viral genome. The VI proteins link the outer capsid shell to the inner icosahedral shell. The VIII proteins help in bonding hexons together and are critical for the stability of the viral capsid. The proteins V, VII, and X are associated with the DNA genome and make up the virion core. Terminal protein binds to each end of the DNA genome [6–8] (**Figure 1**).

The early gene first transcribes E1A protein, an essential protein for viral replication. The E1A protein activates the transcription of other viral genes responsible for viral DNA synthesis. In

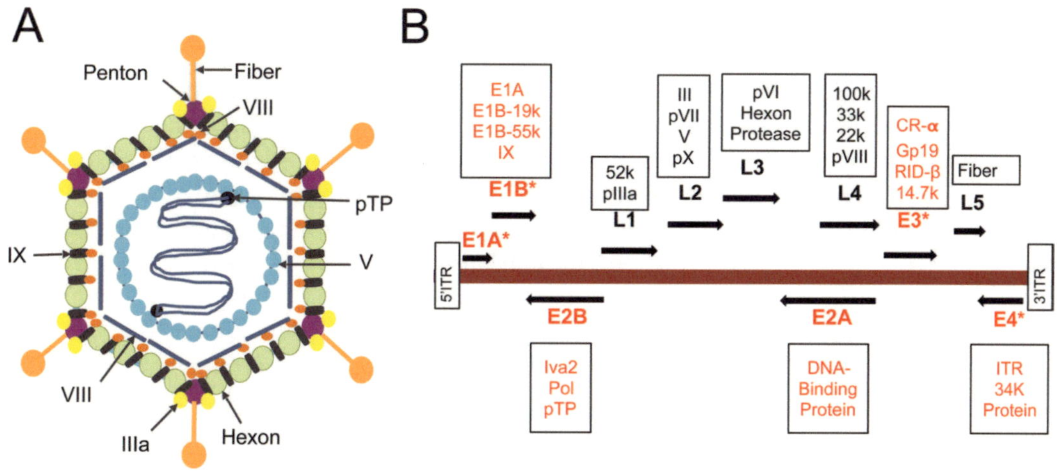

Figure 1. Adenovirus structure and genome organization. (A) Graphical representation of adenovirus structure and various proteins. (B) Adenovirus genome organization showing various early (E) and late (L) transcripts and proteins encoded by each transcript. Regions indicated by red with (*) are deleted in various adenoviral vectors. E1 and E3 regions were deleted in first generation and E1, E2, E3, and/or E4 were deleted in second-generation adenoviral vectors. Most recent adenoviral vectors called helper-dependent adenoviral vectors only contain ITRs and packaging signals. Figure is adapted from Ref. [209].

host cells, E1A stimulates apoptosis by both p53-dependent and -independent pathways [9]. In contrast, the E1B protein inhibits apoptosis by binding to several host cell proteins such as p53, Bak, and BAX proteins [8]. In non-replicating adenoviral vectors, the E1 gene is deleted to render them replication-defective so that it can infect the host cells but cannot multiply. However, for production of non-replicating adenoviral vectors, E1 transfected cells such as HEK293 and PER.C6 are used to allow production of replication-defective adenoviral vector [9].

2.2. Types of adenoviruses

Adenoviruses are grouped under the family Adenoviridae, which is divided into five genera: Mastadenovirus, Aviadenovirus, Siadenovirus, Atadenovirus, and Ichtadenovirus. Human adenoviruses, along with many animal adenoviruses (monkeys, cattle, sheep, swine, dogs), belong to the genus Mastadenovirus. Human adenoviruses (HAd) are classified into seven subgroups: A–G and further in to 67 serotypes based on serological properties. The classification of serotypes into subgroups is based on their similarities in genome organization and DNA sequences, host tropism, carcinogenic potential in rodents, and growth properties in cell cultures. Adenoviral serotyping is based on viral surface antigen neutralizing antibodies and by phylogenetic distance (>10%) in the viral genes that encode viral protease, the protein pVIII, the hexon protein, and the DNA polymerase [10–12].

The genus Aviadenovirus contains bird adenoviruses, while other genera Siadenovirus, Atadenovirus, and Ichtadenovirus contain other adenoviruses of mammals, birds, reptiles, and fishes [13–15]. The adenoviruses isolated from sheep, cattle, deer, possum, and some birds differ from the adenoviruses of the genus Mastadenovirus and are classified under the genus Atadenovirus [6, 16, 17]. The adenoviruses of the genus Mastadenovirus have high A + T (adenine and thymidine)-rich genomes and lack the early region 1 (E1) transcriptional unit. Adenoviruses isolated from many invertebrates are classified under the new genus Siadenovirus. Human and animal adenovirus infections are very common, and the majority of the population of host species contain neutralizing antibodies against the most prevalent serotypes of adenoviruses. Both human and non-human adenoviruses have been studied extensively and are the basis of adenoviral vector-based vaccine and gene therapies [18, 19]. In humans, infection by non-human adenovirus serotypes is not common. However, due to broad tissue tropism and structural and genomic similarity with human adenoviruses, non-human adenoviruses can infect various human tissue types. These properties of adenoviruses encouraged researchers to use non-human adenoviruses as gene or vaccine antigen delivery vectors to mitigate the pre-existing neutralizing immunity that commonly exists against human adenoviral vectors. Several non-human adenoviruses such as bovine Ad serotype 3 (BAd3); canine Ad serotype 2 (CAd2); chimpanzee Ad serotypes 1, 2, 3, 5, 6, 7 and 68 (ChAd1, ChAd2, ChAd3, ChAd5, ChAd6, ChAd7, ChAd68); ovine Ad serotype 7 (OAd7); porcine Ad serotype 3 and 5 (PAd3, PAd5); and fowl Ad serotypes 1, 8, 9, and 10 (FAd1, FAd8, FAd9, FAd10) are currently being tested as vaccine or gene delivery vectors [18, 20–22]. Extensive research in the molecular biology of both human and non-human Ads has helped in better understanding of the adenoviruses and designing of adenoviral vectors.

2.3. Immunity to adenoviruses

Initially, a host detects the invading virus by sensing unique pathogen-associated molecular patterns (PAMPs) present on the pathogen through pattern recognition receptors (PRRs). Once activated, these PRRs transmit signal to express type I interferons (IFNs) and proinflammatory cytokines which inhibit viral replication and recruit various innate immune cells to the site of infection [23–27]. These initial events ensure the efficient activation and presentation of viral antigens by the antigen-presenting cells to T cells and result in the induction of adaptive immune responses. In the following sections, we will discuss innate and adaptive immune responses to Adenoviruses in detail.

2.3.1. Innate immunity

Adenoviruses are known to induce robust innate immune responses in their hosts. The adenovirus binds to its receptor(s) (such as Coxsackie adenovirus receptor or CAR, integrin $\alpha v\beta 5$ heparin sulfate proteoglycans, CD46, sialic acid, etc.) on host cells and gains entry into the cytoplasm [28–32]. However, phagocytic antigen-presenting cells such as macrophages and dendritic cells can also take up virus particles through scavenger receptors [33]. Inside a host cell, the virus can be recognized by various intracellular molecular sensors such as Toll-like receptors (TLRs), RIG-I like receptors (RLRs), nucleotide-binding oligomerization domain (NOD)-like receptors (NLRs), cytosolic DNA sensors, and effector molecules [34–36].

Cytokines such as IL-8 and TNF-α enhance the entry of human adenovirus type C by increasing the availability of CAR and integrin receptors, which facilitate the adenovirus to enter through clathrin-mediated dynamin-dependent endocytosis [27, 33, 37, 38]. The type B human adenoviruses use CD46 or desmoglein-2 and enter host cells through macropinocytosis [39, 40]; this also results in the suppression of IFN-γ-induced production of proinflammatory cytokine IL-12 [41].

One of the major drawbacks of the use of adenovirus in gene therapy is the induction of undesired innate immune responses. In liver and spleen, the resident macrophages can sense and trap blood-borne adenovirus and induce inflammatory response mediators [42, 43]. Adenovirus also activates TLR2-dependent expression of chemokines such as MCP-1 and RANTES. In mice, TLR2 deficiency resulted in reduced NF-κB activation and humoral responses to HAd vector antigens and transgene-encoded antigens [42]. However, TLR2 deficiency did not result in complete inhibition of acute and adaptive responses to HAd, suggesting the involvement of an additional pathway [44]. The cellular β3 integrins were recently reported to interact with arginine-glycine-aspartic acid (RGD) motifs of viral homo-pentameric penton base protein during viral entry, which results in the processing of inactive IL1α into active cytokine in a MyD88-, TRIF-, and TRAF6-independent signaling pathway [43]. The IL1α plays a major role in adenovirus-induced inflammatory responses. The IL1R-deficient mice or wild-type mice treated with anti-IL-1 antibodies demonstrated reduced inflammatory responses as well as hepatotoxicity in adenovirus infection [45]. Further, the interaction between the adenoviral RGD motif and host β3 integrin mediates chemokine secretion, leukocyte infiltration, as well as corneal inflammation in human adenovirus serotype 37 infections [46].

TLR9 also plays a significant role in innate immunity against adenoviruses. Macrophages have been reported to sense adenovirus, helper-dependent adenoviral vector and recombinant E1- and E3-deleted adenovirus through TLR9 [47, 48]. The TLR9-deficient mice show reduced proinflammatory responses and IFN-α production upon adenoviral vector delivery. In a mouse model pf keratitis, adenovirus induced TLR9-dependent IL6 production and monocyte infiltration of the cornea; however, chemokine secretion and keratitis development were TLR9-independent [49, 50]. Another study showed that recombinant adenovirus-induced type I IFN production in plasmacytoid dendritic cells (pDCs) is TLR9-MyD88-dependent but in myeloid DCs (mDCs) and macrophages, it is TLR9-independent [48].

The viral DNA also plays a critical role in the induction of innate immune responses as empty adenoviral particles are found to be poor inducers of innate responses [51]. The presence of double-stranded RNA with 5′-triphosphate groups in the cytoplasm of target cells is sensed by cytosolic PRR such as RIG-I, and viral DNA and RNA are recognized by intracellular PRRs such as TLR3, 7, and 8 present on the endosomal membrane [48, 52–55]. The double-stranded DNA is sensed by TLR9 in the intracellular environment and also by DNA-dependent activator of IRFs (DAI), DNA-dependent protein kinase (DNA-PK), IFN-γ-inducible protein 16 (IFI16), DEAD (Asp-Glu-Ala-Asp) box polypeptide 41 (DDX41), and by cyclic guanosine monophosphate-adenosine monophosphate synthase (cGAS) [34, 35, 56, 57]. Other cytosolic viral DNA sensors are NOD-like receptors (NLRs), which consist of a central nucleotide-binding domain responsible for ATP-dependent self-oligomerization, a C-terminal leucine-rich repeat (LRR) domain that senses the presence of a ligand, and a variable N-terminal interaction domain that mediates protein-protein interactions. The NLR activation leads to the formation of inflammasomes with the help of microtubules [58]. The human adenovirus activates the formation of two types of inflammasomes in myeloid cells: absent in melanoma-2 (AIM2) and NLR-pyrin domain (PYD)-containing protein (NALP3). The activation of inflammasomes induces inflammatory responses via NF-κB signaling, converts pro-IL1β and pro-IL18 into IL1β and IL18, respectively, and can lead to DNA fragmentation, membrane pore formation, and eventually cell death by pyroptosis [59].

The early adenoviral proteins E1, E3, and E4 interfere with the innate immune signaling and help in evading host immune responses. The E1A protein allows hijacking of cell cycle, inducing apoptosis, evading immuning, inducing tumorigenesis, and expressing viral genes [60, 61]. The E1A has been reported to block type 1 IFN-inducible gene expression [62–64], type 2 IFN-γ-dependent HLA class II expression, and IFN-β expression in response to double-stranded RNA due to inhibition of transcription complex formation [62]. Furthermore, E1A inhibits IFN-alpha-stimulated transcription factor 3 (ISGF3), IFN-stimulated genes (ISGs) [65, 66], and immunoproteasomes, resulting in reduced antigen presentation to T cells [67]. In addition, the early adenoviral proteins E1B-19K and E1B-55K antagonize p53-mediated apoptosis [68, 69], and E1B-55K interferes with the induction of IFN-inducible genes [70, 71]. The E1B-55K and E4 proteins induce proteasome-mediated degradation of defense factor death domain-associated protein (Daxx) resulting in the removal of viral transcription blocking allowing viral gene expression [72, 73]. E1B-55K and E4 protein complex also result in inhibition of antiviral innate immune responses [74–77]. The E3 protein has several immune modulatory functions. It blocks the surface transport of MHC-class I molecule and also reduces NK cell receptors on host cells, masking infected cells from detection by immune cells [78]. Furthermore, the E3 protein also inhibits apoptosis of adenovirus-infected cells by downregulation of death receptors [79].

The induction of innate immune responses is critical in adenoviral vector-based strategies. On the one hand, the gene transfer vector should have minimal activation of innate immune signaling to allow efficient gene delivery without immune activation. On the other hand, adenoviral vector-based vaccine antigen delivery could benefit from adenovirus's intrinsic property of innate immune activation that results in efficient activation of transgene-specific adaptive immune responses. Therefore, careful engineering of adenoviral vectors can serve the purpose of both gene and vaccine antigen delivery.

2.3.2. Adaptive immunity

The adaptive immune responses to adenoviruses are directed against both early and late viral proteins. They include both neutralizing antibodies and T cells against viral surface antigens such as hexon, penton, and fiber proteins. However, these adaptive immune responses against adenoviral antigens also present major obstacles in adenoviral vector development as gene delivery, and vaccine antigen carriers limit the number of administrations that can be done and reduce the efficiency of transgene expression. Both humoral and cellular immune responses are discussed in detail in the following sections.

2.3.2.1. Humoral immunity

The surface antigens of adenovirus, penton, hexon, and fiber proteins are involved in host cell receptor interaction and can be neutralized by antibodies. The impact of neutralizing antibodies (nAbs) on adenoviral gene and antigen delivery has been studied extensively [80–82]. The passive transfer of serum from Ad immune mice or purified nAbs against adenoviruses decreases the vector transgene expression and induction of transgene-specific cellular and humoral responses. Depletion of antibodies against fiber, penton, and hexon by affinity chromatography has been shown to significantly enhance the transgene expression and induced immune responses. Furthermore, hexon-specific antibodies seem to play a relatively dominant role *in vivo* in comparison to other antigens. The hexon-specific nAbs are directed against exposed hypervariable loop-containing regions (HVR) on the surface of the virus particle. The pre-existing immunity to a prevalent human adenovirus 5 (HAd5) can be overcome by replacing the entire HAd5 hexon sequence of exposed epitopes with the HVR from a different serotype [6, 82–84]. Consequently, three amino acid substitutions in one of the HVRs significantly reduced neutralization by polyclonal serum raised against a chimpanzee Ad serotype 68 (ChAd68)-derived vector [83, 85]. Conversely, replacing SAd24 hexon with SAd23 into an SAd24/Pan7 vector resulted in reduced transgene expression in mice with pre-existing SAd23/Pan6 immunity [86]. Additionally, nAb against Ad fiber has a minimal effect *in vivo* as shown in a mouse study where nAbs induced by HAd7 administration weakly neutralized HAd7 fiber-expressing chimeric HAd5 vector, indicating that other non-fiber capsid protein-specific nAbs also have a major role in neutralization. Although these *in vivo* animal studies shed light on the relative importance of pre-existing fiber-specific nAbs, they do not accurately reflect the impact of preexisting immunity on adenovector efficiency in humans. In most of these animal studies, the pre-existing fiber-specific nAbs were induced by a single administration of Ad, which induced nAbs with poor breadth and at far lower levels than in humans exposed to repeated natural Ad infections. The nAbs against other surface antigens such as penton may also work against Ad in a synergistic fashion along with fiber-specific nAbs. Together, these factors contribute to more effective neutralization of Ad-based vectors and result in poor transgene expression and induced transgene antigen-specific immunity in humans. Importantly, Ad-specific nAbs are mostly serotype-specific and have very limited to no neutralization capability of other serotypes. This serotype-specificity is due to high sequence heterogeneity of epitopes in the hexon HVR and fiber knob among different Ad serotypes. Moreover, the nAbs are not the only factor in pre-existing Ad immunity; non-neutralizing Abs can also hamper Ad vector efficacy via Fc receptor-dependent cytotoxicity, complement-mediated lysis, and opsonization. In humans, Ad infections are very common and nearly everyone contains some levels of Ad-specific antibodies. The high seroprevalence of Ad-specific antibodies is a major roadblock in adenoviral vector development, and strategies to circumvent these must be examined.

2.3.2.2. Cellular immunity

In humans, Ad vector-specific $CD4^+$ Th1 cells have been detected, but the frequencies of these cells decrease with age [87]. The $CD8^+$ T cell responses to different structural proteins have also been detected in animals in response to adenovirus infection or adenovector administration [19, 80, 88]. Due to extensive homology between different adenoviral structural antigens, both human and mouse-derived $CD4^+$ and $CD8^+$ T cells cross react with human and simian Ad serotypes [20, 89, 90]. Similar to nAb, pre-existing Ad-specific T cells can also reduce Ad vector transgene expression and immunity. Furthermore, Ad-specific T cells have been detected in 80–100% of human subjects in various studies, which make them even more important in Ad vector development [89]. The human studies examining both nAbs and T cells demonstrated a higher proportion of individuals possessing T cell responses compared to nAbs against Ad. The pre-existing Ad-specific T cells have greater consequences for Ad vaccine vector development due to their cross-reactive nature, higher distribution in the human population, and their multifunctional nature [22]. Finally, human Ad vector has also recently been reported to induce cross-reactive hepatitis C virus-specific humoral and cellular immune responses [91]. Widespread use of adenoviral vectors in humans will induce such cross-reactive immune responses at high levels, which might be beneficial or detrimental in the development of natural immunity against HCV and affect the immunopathology and disease progression of HCV infection.

3. Adenoviral vectors

3.1. Construction of adenoviral vectors

Adenoviruses are engineered to make them safe and efficient for human use as vaccine and gene and cancer therapy vectors by deleting certain genome sequences. Initially, human adenoviruses, especially adenovirus 5 (HAd5), were developed as gene delivery vectors. Since the first generation of adenoviral vectors, based on E1 deletion, a number of different strategies have been employed to further improve the gene-carrying capacity and safety by deleting more genes. These strategies can be summarized under following three categories.

3.1.1. First generation

First-generation adenoviral vectors were prepared by deleting early gene E1 to render them replication-defective and create space for a transgene sequence of up to 4.5 kb [92]. Since these vectors lack the E1 region essential for their replication, cell lines such as human embryonic kidney cells (HEK293) were engineered to incorporate viral E1 region [93]. The E1 region in the HEK293 cell line provides trans-complementation and allows the replication of the adenoviral vector [94–97]. These adenoviral vectors carry native tissue transduction capability and efficiently express the transgene in target host cells. However, there are possibilities of spontaneous homologous recombination between vector and E1 regions during amplification inside HEK293 cells, which might enable replication competent adenoviral (RCA) vectors to emerge [98]. To mitigate this problem, another cell line, human embryonic retinoblasts (PERC.6), was made by inserting an expression cassette for the adenoviral E1 region with its own promoter (ubiquitous phosphoglycokinase, PGK) [99]. This eliminates the adenoviral vector homologous regions from the E1 promoter and therefore the chances of recombination [100]. Adenoviral E3 region proteins are known to inhibit immunological pathways [101]. Therefore, the adenoviral E3 region was removed either partially or completely without affecting *in vivo* viral amplification [102]. These deletions in E1 and E3 regions allowed insertion of even larger cargo sequences (up to 8 kb) of two independent genes [103]. Due to the absence of the E1 region, adenoviral vectors are not able to transcribe other early and late viral proteins, although host cellular factors enable these proteins to be expressed at very low levels. This low-level expression of viral protein and subsequent presentation on the cell surface by MHC class I molecules induce robust cytotoxic T cell immune responses. Deletion of E1 is additionally beneficial since adenoviral proteins have toxic effects and induce cell death in a dose-dependent manner [104, 105] (**Figure 2A–C**).

3.1.2. *Second generation*

Second-generation adenoviral vectors possess deletions in E2 or E4 regions that encode for proteins required for replication in target cells [106–108]. These deleted proteins were complemented in trans by cell lines (such as HEK293) to allow for vector propagation. These second-generation vectors provided additional space for larger cargo sequences (10.5 kb) with up to four independent expression cassettes and eliminated the possibility of generating replication-competent adenoviruses during amplification. This deletion of early viral genes impacts the amplification of viral vector in cell culture and results in lower yields due to inefficient complementation by the producer cell lines [107, 109]. These vectors also have been reported to have lower transgene expression. Immunogenicity and cellular toxicity are still a major concern in the second-generation adenoviral vectors [110] (**Figure 2A–C**).

3.1.3. *Third generation*

Third-generation adenoviral vectors are also called "high capacity adenoviral vectors" (HCAds) because they can accept cargo sequences up to 36 Kb [111–113]. The HCAds were generated by deleting all viral sequences except the ITRs and the packaging signal [114]. For replication of third-generation adenovirus vectors in cell culture, instead of the complementation by the viral genes encoded by host cells, an additional adenoviral helper virus is provided. Therefore, the third-generation adenoviral vectors are also called helper-dependent or "gutless" adenoviral vectors [115–117]. The helper adenovirus is generated like a first-generation adenoviral vector and includes packaging signal flanking loxP sites. The vector is produced in HEK293 cells that constitutively express Cre recombinase by simultaneously transducing helper virus and the HCAd genome. This allows the synthesis of adenoviral proteins by the helper virus and enables assembly of viral capsids,

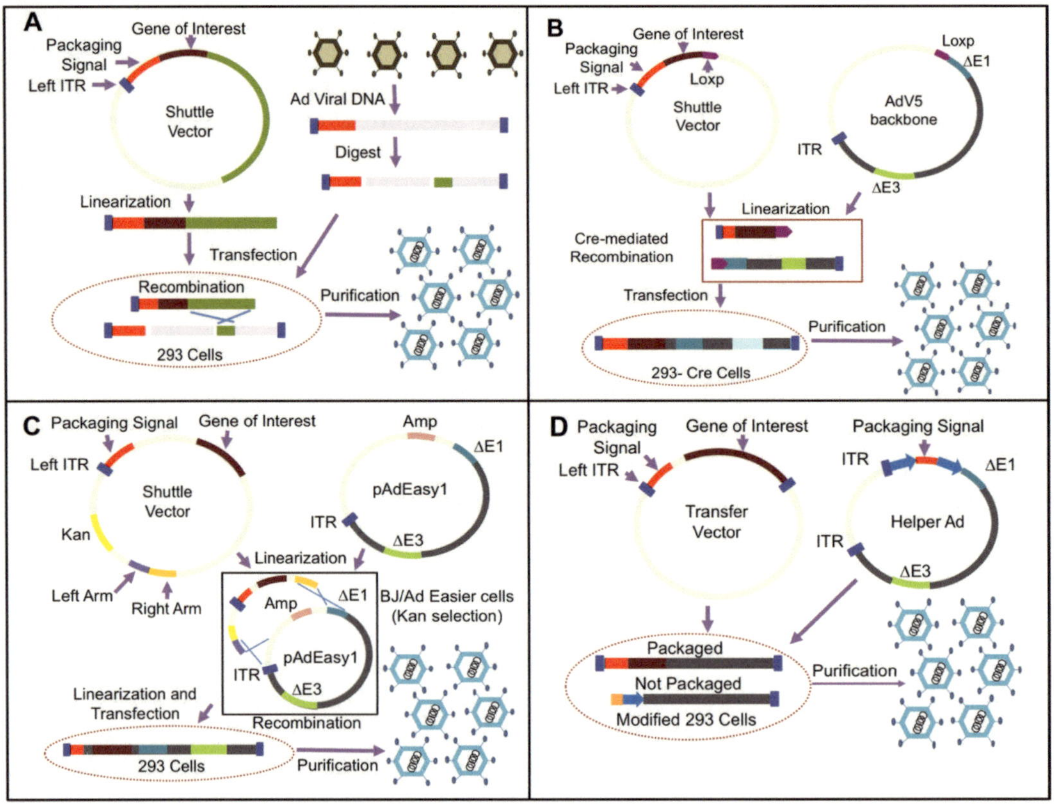

Figure 2. Methods of preparation of different types of adenoviral vectors. (A) First generation. The target gene is cloned into a shuttle vector containing 5′-ITR, a packaging signal, and the sequence for homologous recombination. This shuttle vector and an adenoviral backbone vector are transfected into HEK-293 cells, and adenoviral vector is created through homologous recombination between the two vectors. (B) First or second generation. The target gene is cloned into a shuttle vector that contains 5′-ITR, a packaging signal, and an LoxP site(s). This shuttle vector and a LoxP-containing adenoviral backbone vector are joined together through Cre recombinase-mediated recombination either *in vitro* or in HEK-293 cells. (C) First or second generation. The target gene is cloned into a shuttle vector containing 5′-ITR, a packaging signal, and a kanamycin-containing bacterial replication sequence flanked with two homologous arms. The homologous recombination between the linearized shuttle vector and ampicillin-resistant adenoviral backbone vector takes place in bacterial cells (BJ5183), and adenoviral plasmids are selected on kanamycin. This plasmid is linearized and transfected in HEK-293 cells for adenoviral vector production. (D) Third generation. The target gene is cloned into a transfer vector that only contains ITRs and a packaging signal. A helper adenovirus is used to generate the adenoviral vector. Modified HEK-293 cells are used for adenoviral production, which prevent packaging of helper adenovirus due to deletion of packaging signal. Figure is adapted from Ref. [209].

resulting in the packaging of HCAd genome only. The helper virus genome-packaging signal is excised by Cre-mediated recombination of the loxP sites, thus preventing helper virus genomes from assembling into viral particles. In some production systems, other recombinases like *Saccharomyces cerevisiae*-derived Flp recombinase [118] or bacteriophage-derived phiC31 integrase [119] have also been used. Third-generation vectors have several benefits over first- and second-generation adenoviral vectors. These include less cellular toxicity and reduced immunogenicity [120, 121], thereby providing a flexible vector system that efficiently transduces host cells due to reduced induction of anti-adenoviral neutralizing antibodies [118, 122]. The HCAds can simultaneously encode multiple transgene cassettes. Although the HCAds provide a much superior vector system, they are more complicated to generate compared to previous generations of adenoviral vectors and also have possibility of helper virus contamination due to inefficient Cre-mediated excision of the helper virus packaging signal [123] (**Figure 2D**).

3.2. Current applications of adenoviral vectors

Since their first use in gene therapy, adenoviral vectors have progressed significantly and are currently being tested clinically in several gene therapy, vaccine vector, and anticancer studies.

3.2.1. Gene therapy

Adenoviruses have a unique ability to infect a broad range of cell types. Therefore, adenovirus-based vectors can be used to transduce and deliver transgenes to different cell types including both replicating and quiescent cell populations. This property of adenoviral vectors is extremely important in gene therapy and puts adenoviral vectors on top of viral vectors for gene delivery. Furthermore, adenovirus vectors do not integrate into host genomes but stay as episomal DNA in the nucleus of host cells. Modern adenoviral vectors can take multiple gene cassettes, up to 36 kb of foreign DNA, which make them suitable for delivering virtually any size of gene. In 1992, for the first time, a first-generation adenoviral vector was used to deliver and express alpha-1 antitrypsin (A1AT) in hepatocytes of a patient who had alpha-1 antitrypsin deficiency [124]. In another study, an E1–E3 deleted HAd5 adenoviral vector was used to deliver an A1AT gene to lung tissues [125]. Later, using adenoviral vectors, a number of attempts were made to deliver dysfunctional or deficient genes, which were responsible for several human genetic diseases and conditions. Cystic fibrosis is one such human genetic disease, in which the gene CFTR (cystic fibrosis transmembrane conductance regulator) becomes dysfunctional due to mutation. Adenoviral vector was used to deliver CFTR genes to lung tissues [126]. In another study, the adenoviral vector was used to deliver the gene for ornithine transcarbamylase, which is required in the urea cycle and is responsible for ornithine-transcarbamylase deficiency [127, 128]. These studies faced several challenges including humoral and cellular immunity to adenoviral vectors upon repeated administration of vector, cellular cytotoxicity, and oncogenesis [129]. These trials raised serious safety concerns for using adenoviral vectors in gene therapy and resulted in a sharp decline in their use. The reasons for these problems were studied extensively and addressed by constructing new adenoviral vectors. The adenoviral immunogenicity and cytotoxicity were suspected to be due to low-level expression of several viral proteins. The newer generations of adenoviral vectors had these adenoviral genes removed, and hence the vector immunogenicity and toxicity were significantly reduced. The new generations of adenoviral vectors have raised new hope in adenoviral vector-based gene delivery. Currently, a number of gene therapy clinical trials are ongoing with adenoviral vectors (**Table 1**). The previous generation of adenoviral vectors is still in use for vaccine antigen delivery due to their inherent capability of inducing robust humoral and cellular immune responses.

3.2.2. Vaccine vector

As described earlier, adenoviruses activate several innate immune signaling pathways that result in the secretion of a number of proinflammatory cytokines. These proinflammatory cytokines pave the way for effective immune cell stimulation and result in the induction of robust adaptive humoral and cellular immune responses. To resolve infections with intracellular pathogens such as viruses, $CD8^+$ cytotoxic T lymphocyte (CTL) responses are critical. Transgene antigens carried by adenoviral vectors are presented to T cells via MHC class I molecules, and therefore, they induce efficient and robust CTL responses. The CTLs efficiently recognize and kill virus-infected cells, intracellular pathogens, and cancerous cells. These properties make adenoviral vectors promising as vaccine vectors. A number of human clinical trials have been conducted for adenoviral vector-based vaccines against different infectious diseases including Ebola virus, Zika virus, influenza viruses, HIV, *Mycobacterium tuberculosis*, and malaria [21, 130].

An HAd5 vector-based HIV vaccine containing clade B sequences of *gag/pol* and *nef* genes was tested in several clinical trials by Merck during 2003–2006 [131]. These studies demonstrated that a majority (80%) of vaccine recipients induced T cells with a magnitude of 275–300 IFN-γ producing cells per million peripheral blood mononuclear cells (PBMCs), where about 50% vaccinees had detectable and durable HIV-specific $CD8^+$ T and $CD4^+$ T cells. This is far greater than any other T cell vaccine at that time [132–135]. These results were very encouraging and led to a multinational STEP trial involving about 3000 subjects in 2005 [136, 137]. The early results of this vaccine indicated that vaccine was well

S. no.	Adenoviral vector (biologic)	Modification	Transgene	Target/ condition	Phase	ClinicalTrials identifier
1	HAd5-CB-CFTR	E1 deleted	Cystic fibrosis transmembrane conductance regulator (CFTR) gene	Cystic fibrosis	I	NCT00004779
2	HAd5-hAQP1	E1 deleted	Human aquaporin-1 (hAQP1)	Parotid salivary dysfunction	I	NCT00372320
3	HAd5-PDGF-B	E1 deleted	Platelet-derived growth factor B (PDGF-B)	Varicose ulcer	I	NCT00000431
4	HAd5-PEDF (AdGVPEDF.11D)	E1, E3 and E4 deleted	Pigment epithelium-derived factor (PEDF) protein	Macular degeneration	I	NCT00109499
5	HAd5-VEGF	E1–E3-deleted	Vascular endothelial growth factor D (VEGF-D) gene	Angina pectoris/ myocardial infarction	I	NCT01002430

Table 1. Gene therapy: adenoviral vectors in clinical trial.

tolerated upon repeated vaccination and induced robust T-cell responses to HIV antigens. Despite these early findings, the STEP trial had to be terminated prematurely in 2007 due to enhanced acquisition of HIV infection in the vaccine group compared to placebo [137, 138]. A total of 82 cases of HIV infection were recorded in the trial participants, 49 cases were in vaccine recipients and 33 were in placebo group. Another interesting observation was that the HIV infection rate was twofold higher in men with prior adenovirus type 5 infection (Ad5 titers >18) versus placebo recipients [135, 139]. The same HAd5 clade B *gag/pol* and *nef* genes-based vaccine was tested in another companion Phambili clinical trial in a South African population. The goal was to investigate whether this vaccine would be efficacious against clade C HIV infections. The participants had different prevalent modes of sexual transmission, different subtypes of HIV-1, and varying Ad5 seroprevalence. Unfortunately, this trial also had to be stopped due to acquisition of HIV infection in 9 females seropositive for HAd5 out of a total of 11 cases. Of these 9 cases, 6 were vaccinees [140, 141]. These results indicated that pre-existing immunity to the Ad5 vector is an important risk factor for HIV acquisition among vaccine recipients. Such profound effects of pre-existing Ad5 immunity on HIV acquisition were not observed in previous studies in non-human primates (NHP) using an adenovirus vector-based vaccine [142]. Several hypotheses were provided for the failure of the STEP trial but none of the hypotheses were proven after experimentation, and the mechanisms for higher HIV acquisition in vaccinees with pre-existing Ad immunity still remain unclear. These results shocked the vaccine community and raised serious questions regarding the fate of adenoviral vectors in vaccine approaches.

In 2009, another famous HIV vaccine clinical trial (HVTN phase II) was started [143]. It utilized a heterologous prime-boost strategy, in which vaccinees were first primed with a DNA-based vaccine expressing HIV proteins (envA, envB, envC, gagB, polB, nefB) followed by a booster Ad5 vector vaccine having matching HIV antigens as transgenes. This trial too met with the same fate as STEP, as the vaccine failed to reduce the HIV acquisition rate or attenuate the disease in infected subjects. It was terminated in 2013 prior to completion. However, this study and several other studies provided evidence of superior response rates, induction of broader T cell immune responses with well-accepted tolerance, by heterologous prime-boost vectors compared to homologous vaccination [144]. This started a new vaccination regimen involving priming with one type of adenoviral vector and boosting with another adenoviral vector derived from novel serotypes such as HAd26 and HAd35. This allowed for repeated vaccination and also vaccination in individuals with pre-existing vector immunity [145–149]. Currently, a number of non-human adenoviral vectors such as chimpanzee and bovine are also being utilized to avoid pre-existing vector-specific immunity [18, 21, 150–152].

Due to the emergence of life-threatening infectious diseases such as Ebola and Zika viruses, an immediate need for vaccines for these pathogens was recognized. These urgent needs attracted researchers toward viral vector platform-based vaccines, especially extensively studied and improved vector technology, and adenoviral vectors became the focus of several vaccines against these infectious diseases [21, 130]. Adenoviral vector-based vaccines are easy to design and to produce on a mass scale, which is of paramount significance for clinical use. Therefore, three adenoviral vector-based vaccines encoding Ebola virus glycoprotein, ChAd3-ZEBOV1 from GlaxoSmithKline [153], Ad26-ZEBOV/MVA-BN-Filo2 from Johnson & Johnson [154], and HAd5 from the Chinese federal agency [155] were quickly generated and tested in macaques. Each of them proved to be well tolerated, immunogenic, and protective in macaques. All these vaccines were also well tolerated, safe, and immunogenic in phase I clinical trials, and the ChAd3- and ChAd26/MVA-based vaccines progressed further into phase II and phase III efficacy trials [156, 157]. The Chinese Ad5-based Ebola vaccine showed less efficacy in phase I clinical trial in individuals with pre-existing adenoviral immunity [158]. Beside these, a chimpanzee adenoviral vector ChAd63 prime/MVA boost-based malaria vaccine, which contains *Plasmodium falciparum*-derived ME-TRAP antigen, showed a significant enhancement in antigen-specific T cell responses and partial protection against malarial parasites in a phase I clinical trial [156, 159].

Despite initial setbacks, adenoviral vector-based vaccines are still very attractive and promising vaccine platforms. Currently, several adenoviral vector-based vaccines are in different stages of clinical development (**Table 2**).

3.2.3. Cancer immunotherapy

Several DNA viruses such as reo, measles, herpes simplex, Newcastle disease, and vaccinia have been tested in clinical trials for anticancer immunotherapy. The mechanism of anticancer activity of these viruses is multipronged. One mechanism involves selective infection and replication in tumor cells where expression of viral antigens or oncogenes inside the cancer cell changes the tumor microenvironment by inducing proinflammatory cytokines. These subsequently attract immune cells to the tumors eventually resulting in lysis of the tumor cells. Another mechanism uses vectors to deliver gene(s) whose expression results in the apoptosis of cancer cells or lysis of cancer cells due to replication by replication-competent viral vectors [160, 161]. Adenoviral vector technologies have progressed to the clinical stage for various cancers and also have been approved in some countries for use in human [162–164].

Various anticancer approaches have been tested using adenoviral vectors. One of these approaches depends on the induction of immune responses by delivering specific tumor-associated antigen as a vaccine, which activates immune cells against the tumor [165–168]. Due to immunogenic properties of adenoviral proteins such as capsid, adenoviral vectors induce robust CTL responses, which eventually kill the tumor cells expressing these tumor antigens. However, this vaccination strategy has shown limited success in cancer.

Another approach uses conditional replicative adenoviral vector (CRAd) to preferentially replicate inside a tumor cell and eventually lyse it through a lytic replication [162]. This strategy takes advantage of the conducive nature of cancer cells toward adenoviruses. Adenoviral vectors have been modified to efficiently carry out oncolytic replication in cancer cells while limiting their replication in healthy cells [9, 169]. An adenoviral vector with a partial E1B gene deficiency called ONYX-015 became the first ever adenoviral vector to enter clinical trial in 1996 [169]. The ONYX-015 is unable to replicate in healthy cells expressing p53 but replicates

S. no.	Adenoviral vector (biologic)	Modification	Transgene	Target/ condition	Phase	ClinicalTrials identifier
1	HAd5-EBOV	E1-deleted	Glycoprotein	Ebola virus disease	II	NCT02575456
2	ChAd3-EBO-Z	E1-deleted	Glycoprotein	Ebola virus disease	I/II	NCT02289027
3	ChAd3-EBO-Z/ MVA-EBOV-Z-BN-Filo Prime/Boost	E1-deleted	Glycoprotein/envelope filovirus	Ebola virus disease	I	NCT02267109
4	Ad26-EBOV-Z/MVA-BN-Filo Prime/boost	E1 and E3 deleted	Glycoprotein	Ebola virus disease	I	NCT02891980
5	Ad26-EBOV-Z/MVA-BN-Filo Prime/boost	E1 and E3 deleted	Glycoprotein	Ebola virus disease	III	NCT02661464
6	HAd6-Nsmut/ VhAd3NSmut or CHAd3NSmut/ HAd6NSmut Prime/ boost	E1-deleted	Non-structural protein	Hepatitis C	I	NCT01070407
7	AdCh3NSmut/ Ad6NSmut	E1-deleted	Non-structural protein	Hepatitis C	I	NCT01094873
8	HAd5-HA (VXA-1.1)	E1-deleted	Hemagglutinin and double-stranded RNA as an adjuvant	Influenza H1N1	II	NCT02918006
9	HAd5-HA	E1/E3-deleted	Hemagglutinin	Influenza H5N1	I	NCT00755703
10	HAd4-HA5-Vtn HA	E3-partial deletion	Hemagglutinin	Influenza H5N1	I	NCT01006798
11	HAd35-CSP/HAd26-CSP Vectors Prime/ boost	E1 and E3 deleted	Circumsporozoite (CSP) antigen	Malaria	I/II	NCT01397227
12	HAd5 (NMRC-M3V-Ad-PfCA)	E1, E4 deleted, E3 partially deleted	Circumsporozoite (CSP) antigen, apical membrane antigen 1 (AMA1)	Malaria	I/II	NCT00392015
13	ChAd63-ME-TRAP/ MVA-ME-TRAP poxvirus Prime/boost	E1-deleted	ME-TRAP antigen (pre-erythrocytic thrombospondin-related adhesion protein)	Malaria	I	NCT01373879
14	HAd35-TB Antigens/ MVA85A Prime/boost (AERAS-402, BCG)	E1-deleted, HAd5 E4 orf6 replaced	TB antigens: Ag85A, Ag85B, and TB10.4	Tuberculosis	I	NCT01683773
15	HAd35-TB Antigens (AERAS-402)	E1-deleted, HAd5 E4 orf6 replaced	TB antigens: Ag85A, Ag85B, and TB10.4	Tuberculosis	II	NCT02414828

S. no.	Adenoviral vector (biologic)	Modification	Transgene	Target/ condition	Phase	ClinicalTrials identifier
16	Four HAd5 vectors for four HIVAntigens (VRC-HIVADV014-00-VP/VRC-HIVADV014-00-VP)	E1, E4 and partial E3 deleted	HIV antigens gp140(A), gp140(B) dv12, gp140(C) and GagPol(B)	HIV infections	I	NCT01549509, NCT00119873, NCT00091416, NCT00709605, NCT00102089
17	Plasmid DNA expressing Gag, Plo and Nef +Four HAd5 vectors for four HIVAntigens (VRC-HIVDNA016-00-VP/ VRC-HIVADV014-00-VP), DNA + HAd5/ HAd5 Prime/boost	E1, E4 and partial E3 deleted	HIV antigens gp140(A), gp140(B) dv12, gp140(C) and GagPol(B)	HIV infections	I/II	NCT00123968, NCT00125970
18	Plasmid DNA vaccine/ HAd5-HIV-1 (VRC-HIVDNA016-00-VP/ VRC-HIVADV014-00-VP) prime/boost	E1, E4 and partial E3 deleted	HIV antigens gp140(A), gp140(B) dv12, gp140(C) and GagPol(B)	HIV infections	II	NCT00865566
19	DNA Vaccine/ HAd35/HAd5 (VRC-HIVDNA044-00-VP/ VRC-HIVADV027-00-VP/VRC-HIVADV038-00-VP) HAd35/HAd5 prime/ boost or DNA/HAd5 prime/boost or DNA/ HAd35 prime/boost	E1-deleted	Gag, pol and Nef antigens	HIV infections	I	NCT00801697
20	HAd26 (HAd26. ENVA.01)	E1 and E3 deleted	Env gp140	HIV infections	I	NCT00618605, NCT01103687
21	HAd4-mgag, HAd4-EnvC150 alone or combination	Replication competent	mosaic HIV Gag antigen, HIV clade C Env protein (gp150 1086.C)	HIV infections	I	NCT02771730, NCT01989533
22	rcAd26.MOS1.HIV-Env 1	E3 or E3/E4 deleted	HIV-1 Mos1Env	HIV infections	I	NCT02366013
23	Ad26.Mos.HIV (Ad26. Mos.1.Env + Ad26. Mos1.Gag-Pol + Ad26. Mos2.Gag-Pol)/MVA-Mosaic prime/boost	E3 or E3/E4 deleted	Hiv gag, pol and Env	HIV infections	I/II	NCT02919306

Table 2. Vaccine delivery: adenoviral vectors in clinical trials.

in p53-deficient tumor cells and results in the lysis of the cell, taking advantage of the cancer cell environment that supports vector replication [9]. The ONYX-015 has been proven to be safe and well tolerated in patients with various advanced cancers and is reported to be even more effective when administered in combination with standard chemotherapy [169, 170]. A similar adenoviral vector named Oncorine or H101, developed by Shanghai Sunway Biotech, was approved by the Chinese Food and Drug Administration agency for the treatment of head and neck cancer [171, 172]. To further enhance the efficacy, potency, and specificity of the oncolytic adenoviral vectors, a new generation of adenoviral vectors is being tested. These new adenoviral vector systems carry a suicide gene like HSV thymidine kinase or a cytotoxic prodrug under the control of tumor gene/antigen promotor like prostate antigen promotor [173–175].

In some studies, replication-deficient or replication-competent adenoviral vectors were used to deliver transgenes, which express a tumor suppressor protein or cytotoxic/suicide protein that induces cell cycle arrest or a death cascade [176, 177]. More than 50% of cancers have a mutation in tumor suppressor gene p53. Advexin is a replication-deficient adenoviral vector that expresses p53 through a CMV promotor. It was tested in both preclinical and more than a dozen phase I/II clinical trials and proved to be well tolerated and efficacious against colorectal cancer, hepatocellular carcinoma (HCC), non-small cell lung cancer (NSCLC), prostate cancer, breast cancer, ovarian cancer, bladder cancer, glioma, and squamous cell carcinoma of the head and neck [171, 178, 179]. Gendicine, a similar adenovirus vector developed by a Chinese Biotech Company, Shenzhen SiBiono GeneTech, differs only in that its transgene promotor is from Rous Sarcoma Virus. In 2003, Gendicine was approved by the Chinese Food and Drug Administration agency as a first-ever gene therapy product to be used in combination with chemotherapy to treat head and neck squamous cell carcinoma [4, 178, 180]. Since then, Gendicine has been tested in a number of clinical trials against different types of cancers such as HCC, NSCLC, malignant glioma, and epithelial ovarian carcinoma. It is reported to be well tolerated and provide progression-free long-term survival benefits in combination regimens when compared to standard therapies alone [4, 178, 180, 181]. Therefore, adenoviral vectors have been clinically successful in anticancer therapy and have shown tremendous potential in the treatment of several cancer types [4, 171, 180, 181]. However, there is still scope for further improvements in clinical efficacy and safety (**Table 3**).

3.3. Challenges and solutions to adenoviral vectors use

Development of adenoviral vectors has come a long way since their first use. Currently, different types of adenoviral vectors are available for different applications. In the beginning, a high prevalence of pre-existing immunity to adenoviral vectors was considered as a serious concern for their use in mass vaccination and gene therapeutic applications. Further, immunogenicity, cellular toxicity, and oncogenesis were also major obstacles in gene therapy applications. Many other concerns, such as the possibility of vectors regaining replication competence, non-specificity, immunodominance of adenoviral antigens over the vaccine transgene antigen(s), immune modulation by viral antigens, heterologous immunity with other pathogens, are still evident in many adenoviral vector-based vaccine and gene therapy approaches [182]. These are discussed in the following sections.

S. no.	Adenoviral vector (biologic)	Modification	Transgene	Target/condition	Phase	ClinicalTrials identifier
1	ICOVIR-5	E2F-E1A Δ24 RGD	—	Solid tumors	I	NCT01864759
2	LOAd703	5/3 Δ24	CD40L & 4-1BBL	Pancreatic cancer	I/IIa	NCT02705196
3	HAd5-yCD/ mutTKSR39rep-hIL12	E1B-55K	Cytosine deaminase (CD)/ tyrosine kinase (TK) hIL12	Prostate cancer	I	NCT02555397
4	ONCOS-102 with cyclophosphamide	5/3 Δ24	GM-CSF	Advanced neoplasms	I	NCT01598129
5	VCN-01 with or without abraxane and gemcitabine	DM-1-E2F-E1A Δ24 RGD	Hyaluronidase	Advanced solid tumors	I	NCT02045602
6	VCN-01 with abraxane and gemcitabine	DM-1-E2F-E1A Δ24 RGD	Hyaluronidase	Advanced pancreatic cancer	I	NCT02045589
7	CG0070	E2F-E1A	Granulocyte macrophage colony-stimulating factor (GM-CSF)	Bladder cancer	III	NCT02365818
8	CG0070	E2F-E1A	GM-CSF	Bladder cancer	II/III	NCT01438112
9	Colo-Ad1	Ad11p/Ad3	—	Colon, non-small cell lung cancer, bladder, renal cancer	I	NCT02053220
10	DNX-2401 with Temozolomide	E1A Δ24 RGD	—	Glioblastoma multiforme	I	NCT01956734
11	DNX-2401 with IFNγ	E1A Δ24 RGD	—	Brain tumors	I	NCT02197169
12	Ad5-yCD/mutTKSR39rep-ADP with intensity-modulated radiation therapy (IMRT)	E1B-55K	CD/TK	Prostate carcinoma	II/III	NCT00583492
13	OBP-301	hTERT	—	Hepatocellular carcinoma	I/II	NCT02293850

Table 3. Oncolytic therapy: adenoviruses in clinical trial.

3.3.1. Tissue tropism and transgene expression

Adenoviruses can infect a diverse range of mammalian cell types. Infection to host cells is mediated by binding of adenoviral fiber protein to host cell surface receptors followed by recruitment of RGD motifs on penton bases to bind the host cell alpha-integrins. Most common human adenoviruses (HAd5 and HAd2) bind to coxsackie adenovirus receptor (CAR) present on many different cell types including epithelial, endothelial, hepatocytes, myoblasts, and heart muscle cells. Some cells such as lymphocytes do not express CAR themselves but harbor CAR-recognizing adenoviruses. Adenovirus from subgroup B such as HAd35 do not bind to CAR but recognize another complement regulatory receptor CD46 present on most nucleated human cells, hematopoietic stem cells, and dendritic cells. Another subgroup B adenovirus HAd3 binds to CD80 and CD86 costimulatory molecules on antigen-presenting cells [8, 166, 183, 184]. Other cellular receptors such as integrin $\alpha v\beta 5$, heparin sulfate proteoglycans, and sialic acid have also been reported to aid adenoviral entry to the cells [30–32, 185].

3.3.2. *Pre-existing adenoviral immunity*

The impact of pre-existing immunity against adenoviral vectors has been discussed in the previous section under adaptive immunity. To avoid the pre-existing immunity against adenoviral vectors, several strategies are being employed as discussed below.

3.3.2.1. *Use of alternative less frequent adenoviruses for vector*

Several human adenoviruses with low seroprevalence such as HAd2, HAd26, and HAd35 were identified and developed into vectors [186]. The seroprevalence of these rare human adenovirus serotypes is very low, and hence the effect of pre-existing immunity is minimal [81, 187]. HAd26 and HAd35 vectors have been tested in phase I clinical trials and proven to be safe. However, the immunogenicity and efficacy of these low seroprevalence vectors are reported to be lower in comparison to more prevalent HAd5. These results are very concerning and warrant further investigation to find the reasons for the poor performance of these vectors. The nAb and T cells against HAd5 do not cross react with HAd35 but nAbs and T cells against another common serotype, HAd2, cross react with HAd35 and reduce the immunogenicity and efficacy of the HAd35 vectors [188, 189].

To avoid the cross-reactive immunity due to closely related serotypes of Ad, more genetically distant Ad serotypes including animal and bird Ad were developed as viral vectors. Among non-human adenovirus vectors, chimpanzee-derived adenovirus vector (ChAd) is the most widely used. In comparison to HAd, the nAbs against ChAd have been found to be less prevalent. For example, nAbs against ChAd7 were detected in only 15% of American, European, Chinese, and African population [186]. Similarly, nAbs against ChAd6 are also low in these populations except Africans, which have about 40% ChAd6-specific nAbs [81, 186]. Several chimpanzee adenoviral vector-based vaccines, such as ChAd7 for Ebola virus, ChAd6 for rabies, and ChAd6, ChAd7, and ChAd9 for malaria, have shown high efficacy in animal models [190]. Furthermore, ChAd63-based malaria and ChAd3-based hepatitis C virus vaccines have shown to be safe and highly immunogenic in phase I clinical trial [191, 192]. Despite low seroprevalence of ChAd vectors in humans, pre-existing cross-reactive T cells against many conserved viral antigens are still a major concern. The HAd-induced ChAd cross-reactive T cells have been reported against ChAd6, ChAd7, ChAd24, ChAd32, and ChAd68 [22, 80, 89, 90]. The negative effects of these cross-reactive T cells on ChAd have been demonstrated in several animal models [22, 89]. The impact of the pre-existing cross-reactive T cells could be far greater in the clinical setting where humans are repeatedly exposed to various serotypes of adenovirus and carry a far broader diversity and a higher frequency of cross-reactive T cells.

Apart from rare human and chimpanzee adenoviruses, several other adenoviruses derived from animals such as bovine, porcine, ovine, canine, and fowl are also being explored for vector development [151, 193–195]. The human population lack nAbs against these adenoviruses, and therefore, the vectors derived from these adenoviruses could be more efficacious in comparison to HAd and ChAd. The mouse models with experimentally induced pre-existing immunity by the administration of HAd vector demonstrated a lack of nAbs and $CD4^+$ T cells against PAd3 and BAd3 [18]. Furthermore, BAd3- or PAd3-based influenza virus vaccine demonstrated high efficacy even in the presence of pre-existing HAd5 immunity. There was also no effect of pre-existing HAd5 immunity on transgene expression, immunogenicity, and efficacy in animal models. However, their potential benefits still need to be proven in humans.

3.3.2.2. Routes of immunization

Several studies have reported that different routes of immunizations can negate the detrimental effect of pre-existing immunity. This outcome could be at least partly due to evasion of tissue-resident Ad-specific T cells when using different routes of immunization. Tissue-resident CD8 memory T cells remain confined to a specific tissue. As such, they are not systemic and do not prevent Ad vector infection in distant tissues. These tissue-resident T cells are induced by tissue-derived migratory dendritic cells during priming, which activate T cells with specific tissue-homing molecules [196–198]. In a non-human primate model, HAd5-induced protective immune responses by intranasal/intratracheal immunization were not affected by pre-existing HAd5 immunity that had been induced by intramuscular administration of an unrelated HAd5 vector [199]. Although this strategy has shown promising results in animal models, it has not been verified in humans. This strategy is however restricted by limited routes feasible for administration in humans. Moreover, the nAbs induced by adenoviral infection or adenoviral vector-based vaccination can be detected in any part of the body, which may affect the transgene expression and immune responses irrespective of the alternative route used for subsequent adenovector administration. However, nAbs are generally present in blood and are in most cases not cross-reactive to different serotypes or subtypes and Ads from different host species [200].

3.3.2.3. Heterologous prime-boost strategy

Another strategy to avoid pre-existing adenovector immunity involves the use of heterologous prime-boost regimens. In this strategy, the priming and boosting are done by using different antigen delivery vehicle and/or vectors derived from either different serotypes of the same species or vectors from completely different host species, for example, priming with DNA and boosting with HAd5 or priming with ChAd68 and boosting with ChAd1 [201]. Studies have shown that the heterologous prime-boost induces more robust immune responses compared to single vaccination or homologous prime-boost immunizations. The cellular immune responses induced by DNA prime and HAd boost were not affected by pre-existing HAd5 immunity. These findings were further confirmed in a clinical trial. Another preclinical study involving *Plasmodium* or SARS antigens encoded by Modified Vaccinia Ankara (MVA)/adenoviral vector as prime/boost showed induction of robust T cell and Ab responses of higher magnitude compared to Ad/DNA regimens. Finally, a ChAd63/MVA prime-boost strategy is now being evaluated in clinical trials as a vaccine against malaria, HIV, and HCV [191, 202–204]. The heterologous prime/boost strategy seems very effective in circumventing the pre-existing immunity against adenoviral vectors in studies conducted so far; however, the vehicle for priming and/or boosting must be carefully designed and selected.

3.3.3. Immunodominance over transgene immunity

Adenoviral antigens induce robust antibody and T cell immune responses. Recent studies have shown that adenoviral-derived epitopes can dominate over the transgene-derived epitopes and hinder the induction of transgene-specific immunity. This impairment of transgene-specific immune responses in naive vaccinees is due to immune competition. Epitopes derived from an adenovirus vector were shown to inhibit the induction of HIV GagL85-93-specific $CD8^+$ T cells [205]. This study demonstrated that competition occurs at the level of responding $CD8^+$ T cells, and co-immunization with an interleukin 2-encoding plasmid restored GagL85-93-specific $CD8^+$ T cell responses in the presence of an adenoviral hexon486-494 epitope. The IL2, however, could not restore GagL85-93 responsiveness in Ad-based immunization, likely due to the presence of other epitopes in the Ad vector [206]. Another study demonstrated that plasmid DNA, but not adenovirus vector-encoding hepatitis B surface antigen (HBsAg), primed CD8 T cells against subdominant HBsAg epitopes [207]. These studies suggest that adenoviral antigen-specific T-cell immunity is primed efficiently during adenoviral vector-based immunization, which can limit the immunogenicity of adenoviral vector-encoded transgenic antigens. These studies highlight the need for modifications of the vector or the transgene used in immunization to circumvent or dominate over the adenoviral vector-specific epitopes and induce more effective transgene-specific immunity.

3.3.4. Heterologous immunity induced by adenoviral antigens

We discovered an unusual and interesting phenomenon that non-recombinant HAd5 vector induces robust cross-reactive immune responses toward hepatitis C virus (HCV) antigens [91]. Upon further investigation, we found that adenoviral proteins contain extensive homologies with various peptide epitopes derived from HCV antigens. These observations led us to investigate the adenoviral vector-induced HCV cross-reactive immune responses in detail in both mice and humans. In mice, we demonstrated that Ad vector alone can induce potent, broad anti-HCV cross-reactive immunity that can significantly reduce viral load upon challenge with infectious chimeric Vaccinia-HCV. Furthermore, we also detected

HCV cross-reactive antibodies and HCV antigen-dependent expression of IFN-γ in T cells from a cohort of HCV-naïve but Ad-immune human individuals. Previous studies have also reported that one pathogen can induce cross-reactive immunity against an unrelated pathogen [91, 208]. This kind of immunity is known as heterologous immunity. Heterologous immunity is a double-edged sword, which can modulate the breadth of the T cell repertoire, influence the memory T cell pool and/or the immune dominance of a specific epitope, and lead to enhanced or diminished immune responses against a pathogen. These observations have significant clinical implications on natural history, immunopathogenesis, and disease outcome in HCV infection. The widespread use of adenoviral vectors in mass vaccination programs might change the immune hierarchy and natural T cell responses against HCV antigens and deviate and/or alter the incidence of HCV infection and immune pathogenesis in an at-risk population. However, a careful evaluation of adenoviral-induced cross-reactive immune responses and their impact on HCV immunity and immunopathology is needed to more accurately ascertain the impact of this phenomenon.

4. Conclusions and future prospects

Since their first use as gene delivery vehicles, adenoviral vectors have been extensively studied in a number of applications and have improved substantially over time. Issues such as toxicity, pre-existing immunity in humans, and challenges in construction are continually being addressed. More recently, outbreaks of newly emerging infectious diseases, such as SARS, Ebola, and Zika, and the continuing threat of bioterrorism have increased the requirement of novel vaccine platforms which can be designed and produced in large scale within a short period of time. Adenoviral vectors, due to their versatility, ease of construction, adeptness to rapid mass production, and induction of robust transgene-specific humoral and cellular immune responses, have proven to be valuable in the development of vaccines for emerging viral infectious diseases. Furthermore, extensive knowledge about the adenoviral vector biology and the induced immune responses in animals have tremendously helped in developing effective vaccine candidates against several viral pathogens, which have progressed to advanced clinical stages. The adenoviruses are now at the forefront of vaccinology and have shown huge potential in both pre-clinical and clinical studies for HIV, malaria, Ebola virus, and Zika virus vaccines. Apart from infectious disease, adenoviral vectors have been approved for human use as cancer and gene therapy. Therefore, adenoviral vectors have opened new avenues in gene delivery, vaccine antigen delivery, and cancer molecular therapy.

Acknowledgements

We acknowledge Bharti Singh for designing and creating figures for this chapter. This work was supported by the Canadian Institutes of Health Research (CIHR) operating grant to BA (MOP 79327).

Conflict of interest

The authors SS, BA, and RK are co-inventors of a patent on adenoviral vector-based method of inducing immune responses against hepatitis C virus.

Author details

Shakti Singh[1,3], Rakesh Kumar[2] and Babita Agrawal[3]*

*Address all correspondence to: bagrawal@ualberta.ca

1 Division of Infectious Disease, Los Angeles Biomedical Research Institute, Harbor-University of California Medical Center, Torrance, California, USA

2 Department of Laboratory Medicine and Pathology, Faculty of Medicine and Dentistry, University of Alberta, Edmonton, Alberta, Canada

3 Department of Surgery, Faculty of Medicine and Dentistry, University of Alberta, Edmonton, Alberta, Canada

References

[1] Wold W, Horwitz M. Adenoviruses. In: Fields BN, Knipe DM, Howley PM, editors. Fields Virology. 6th ed. 5th ed. Philadelphia: Lippincott-Raven; 2007. pp. 2395-2436

[2] Rowe W, Hueer R, Gilmore L, Parrot R, Ward T. Isolation of a cytopathogenic agent from human adenoids undergoing spontaneous degeneration in tissue culture. Proceedings of the Society for Experimental Biology and Medicine Society for Experimental Biology and Medicine (New York, NY). 1953;**84**(3):570-573

[3] Yu B, Wang Z, Dong J, Wang C, Gu L, Sun C, et al. A serological survey of human adenovirus serotype 2 and 5 circulating pediatric populations in Changchun, China, 2011. Virology Journal. 2012;**9**:287

[4] Zhang W, Li L, Li D, Liu J, Li X, Li W, et al. The first approved gene therapy product for cancer Ad-p53 (gendicine): 12 years in the clinic. Human Gene Therapy. 2018; **29**(2):160-179

[5] McConnell M, Imperiale MJ. Biology of adenovirus and its use as a vector for gene therapy. Human Gene Therapy. 2004;**15**(11):1022-1033

[6] Crawford-Miksza L, Schnurr D. Analysis of 15 adenovirus hexon proteins reveals the location and structure of seven hypervariable regions containing serotype-specific residues. Journal of Virology. 1996;**70**(3):1836-1844

[7] CMM F, Rosa-Calatrava M, Conway J, Zubieta C, Cusack S, Ruigrok R, et al. A quasi-atomic model of human adenovirus type 5 capsid. The EMBO Journal. 2005;**24**(9):1645-1654

[8] Wickham T, Carrion M, Kovesdi I. Targeting of adenovirus penton base to new receptors through replacement of its RGD motif with other receptor-specific peptide motifs. Gene Therapy. 1995;**2**(10):750-756

[9] Bischoff J, Kirn D, Williams A, Heise C, Horn S, Muna M, et al. An adenovirus mutant that replicates selectively in p53-deficient human tumor cells. Science (New York, NY). 1996;**274**(5286):373-376

[10] Benkö M, Virus-Host H. Molecular evolution of adenoviruses. Current Topics in Microbiology and Immunology. 2003;**272**:3-35

[11] Wadell G. Adenoviridae: The adenoviruses. In: Laboratory Diagnosis of Infectious Diseases Principles and Practice. New York, NY: Springer; 1988

[12] Ginsberg HS. Identification and classification of adenoviruses. Virology. 1962;**18**:312-319

[13] Mase M, Mitake H, Inoue T, Imada T. Identification of group I-III avian adenovirus by PCR coupled with direct sequencing of the hexon gene. The Journal of Veterinary Medical Science. 2009;**71**(9):1239-1242

[14] Calnek B, Shek W, Menendez N, Stiube P. Serological cross-reactivity of avian adenovirus serotypes in an enzyme-linked immunosorbent assay. Avian Diseases. 1982;**26**(4):897-906

[15] Kawamura H, Shimizu F, Tsubahara H. Avian adenovirus: Its properties and serological classification. National Institute of Animal Health Quarterly. 1964;**4**:183-193

[16] Bürki F, Wege H, Reich-Rohrwig C, Hinaidy B. Bovine adenoviruses. I. Characterization and serological classification as types 4 of two isolates from latently infected calf testicles. Zentralblatt für Veterinärmedizin. Reihe B. 1978;**25**(7):555-565

[17] Pereira HG, Huebner RJ, Ginsberg HS, Van Der Veen J. A short description of the adenovirus group. Virology. 1963;**20**:613-620

[18] Guo J, Mondal M, Zhou D. Development of novel vaccine vectors: Chimpanzee adenoviral vectors. Human Vaccines & Immunotherapeutics. 2018;**14**(7):1679-1685

[19] Cheng C, Wang L, Ko S, Kong W, Schmidt S, Gall J, et al. Combination recombinant simian or chimpanzee adenoviral vectors for vaccine development. Vaccine. 2015;**33**(51): 7344-7351

[20] Colloca S, Barnes E, Folgori A, Ammendola V, Capone S, Cirillo A, et al. Vaccine vectors derived from a large collection of simian adenoviruses induce potent cellular immunity across multiple species. Science Translational Medicine. 2012;**4**(115):115ra2-115ra2

[21] Ewer K, Sebastian S, Spencer A, Gilbert S, Hill A, Lambe T. Chimpanzee adenoviral vectors as vaccines for outbreak pathogens. Human Vaccines & Immunotherapeutics. 2017;**13**(12):3020-3032

[22] Iampietro M, Larocca R, Provine N, Abbink P, Kang Z, Bricault C, et al. Immunogenicity and cross-reactivity of rhesus adenoviral vectors. Journal of Virology. 2018;**92**(11): e00159-e00118

[23] Suomalainen M, Nakano MY, Boucke K, Keller S, Greber UF. Adenovirus-activated PKA and p38/MAPK pathways boost microtubule-mediated nuclear targeting of virus. The EMBO Journal. 2001;**20**(6):1310-1319

[24] Tibbles LA, Spurrell JC, Bowen GP, Liu Q, Lam M, Zaiss AK, Robbins SM, Hollenberg MD, Wickham TJ, Muruve DA. Activation of p38 and ERK signaling during adenovirus vector cell entry lead to expression of the CXC chemokine IP-10. Journal of Virology. 2002;**76**(4):1559-1568

[25] Basner-Tschakarjan E, Gaffal E, O'Keeffe M, Tormo D, Limmer A, Wagner H, Hochrein H, Tüting T. Adenovirus efficiently transduces plasmacytoid dendritic cells resulting in TLR9-dependent maturation and IFN-α production. The Journal of Gene Medicine. 2006;**8**(11):1300-1306

[26] Hartman Z, Appledorn D, Amalfitano A. Adenovirus vector induced innate immune responses: Impact upon efficacy and toxicity in gene therapy and vaccine applications. Virus Research. 2008;**132**(1-2):1-14

[27] Lütschg V, Boucke K, Hemmi S, Greber UF. Chemotactic antiviral cytokines promote infectious apical entry of human adenovirus into polarized epithelial cells. Nature Communications. 2011;**2**:391

[28] Amberg N. Adenovirus receptors: Implications for targeting of viral vectors. Trends in Pharmacological Sciences. 2012;**33**(8):442-448

[29] Wolfrum N, Greber UF. Adenovirus signalling in entry. Cellular Microbiology. 2013; **15**(1):53-62

[30] Lyle C, McCormick F. Integrin $\alpha v\beta 5$ is a primary receptor for adenovirus in CAR-negative cells. Virology Journal. 2010;**7**(1):1-13

[31] Lenman A, Liaci MA, Liu Y, Årdahl C, Rajan A, Nilsson E, et al. Human adenovirus 52 uses sialic acid-containing glycoproteins and the coxsackie and adenovirus receptor for binding to target cells. PLOS Pathogens. 2015;**11**(2):e1004657

[32] Arnberg N, Pring-Åkerblom P, Wadell G. Adenovirus type 37 uses sialic acid as a cellular receptor on Chang C cells. Journal of Virology. 2002;**76**(17):8834-8841

[33] Meier O, Gastaldelli M, Boucke K, Hemmi S, Greber UF. Early steps of clathrin-mediated endocytosis involved in phagosomal escape of Fcγ receptor-targeted adenovirus. Journal of Virology. 2005;**79**(4):2604-2613

[34] Rathinam VA, Fitzgerald KA. Cytosolic surveillance and antiviral immunity. Current Opinion in Virology. December 2011;**1**(6):455-462

[35] Rathinam V, Fitzgerald KA. Innate immune sensing of DNA viruses. Virology. 2011; **411**(2):153-162

[36] Takeuchi O, Akira S. Innate immunity to virus infection. Immunological Reviews. 2009; **227**(1):75-86

[37] Wang X, Bergelson JM. Coxsackievirus and adenovirus receptor cytoplasmic and transmembrane domains are not essential for coxsackievirus and adenovirus infection. Journal of Virology. 1999;**73**(3):2559-2562

[38] Wang K, Huang S, Kapoor-Munshi A, Nemerow G. Adenovirus internalization and infection require dynamin. 1998. Journal of Virology. 1998;**72**(4):3455-3458

[39] Gaggar A, Shayakhmetov DM, Lieber A. CD46 is a cellular receptor for group B adenoviruses. 2003. Nature Medicine. 2003;**9**(11):1408-1412

[40] Amstutz B, Gastaldelli M, Kälin S, Imelli N, Boucke K, Wandeler E, Mercer J, Hemmi S, Greber UF. Subversion of CtBP1-controlled macropinocytosis by human adenovirus serotype 3. The EMBO Journal. 2008;**27**(7):956-969

[41] Iacobelli-Martinez M, Nepomuceno RR, Connolly J, Nemerow GR. CD46-utilizing adenoviruses inhibit C/EBPβ-dependent expression of proinflammatory cytokines. Journal of Virology. 2005;**79**(17):11259-11268

[42] Appledorn DM, Patial S, McBride A, Godbehere S, Van Rooijen N, Parameswaran N, Amalfitano A. Adenovirus vector-induced innate inflammatory mediators, MAPK signaling, as well as adaptive immune responses are dependent upon both TLR2 and TLR9 in vivo. 2008. Journal of Immunology. 2008;**181**(3):2134-2144

[43] Di Paolo NC, Miao EA, Iwakura Y, Murali-Krishna K, Aderem A, Flavell RA, Papayannopoulou T, Shayakhmetov DM. Virus binding to a plasma membrane receptor triggers interleukin-1α-mediated proinflammatory macrophage response in vivo. Immunity. 2009;**31**(1):110-121

[44] Fejer G, Drechsel L, Liese J, Schleicher U, Ruzsics Z, Imelli N, et al. Key role of splenic myeloid DCs in the IFN-αβ response to adenoviruses in vivo. PLoS Pathogens. 2008;**4**(11): e1000208

[45] Di Paolo NC, van Rooijen N, Shayakhmetov DM. Redundant and synergistic mechanisms control the sequestration of blood-born adenovirus in the liver. Molecular Therapy. 2009;**17**(4):675-684

[46] Chintakuntlawar A, Astley R, Chodosh J. Adenovirus type 37 keratitis in the C57BL/6J mouse. Investigative Ophthalmology & Visual Science. 2007;**48**(2):781-788

[47] Cerullo V, Seiler M, Clarke C, Erez A, Barry M, Lee B. 150. Metabolically biotinylated helper dependent adenovirus: A new and rapid approach for targeting of high-capacity adenoviral vector. Molecular Therapy. 2006;**13**:S59

[48] Zhu J, Huang X, Yang Y. Innate immune response to adenoviral vectors is mediated by both Toll-like receptor-dependent and -independent pathways. Journal of Virology. 2007;**81**(7):3170-3180

[49] Chintakuntlawar A, Chodosh J. Chemokine CXCL1/KC and its receptor CXCR2 are responsible for neutrophil chemotaxis in adenoviral keratitis. Journal of Interferon & Cytokine Research: The Official Journal of the International Society for Interferon and Cytokine Research. 2009;**29**(10):657-666

[50] Chintakuntlawar A, Zhou X, Rajaiya J, Chodosh J. Viral capsid is a pathogen-associated molecular pattern in adenovirus keratitis. PLoS Pathogens. 2010:e1000841

[51] Iacobelli-Martinez M, Nemerow GR. Preferential activation of Toll-like receptor nine by CD46-utilizing adenoviruses. Journal of Virology. 2007;**81**(3):1305-1312

[52] Weber C, Armbruster N, Scheller C, Kreppel F, Kochanek S, Rethwilm A, et al. Foamy virus-adenovirus hybrid vectors for gene therapy of the arthritides. The Journal of Gene Medicine. 2013;**15**(3-4):155-167

[53] Johnsen IB, Nguyen TT, Ringdal M, Tryggestad AM, Bakke O, Lien E, Espevik T, Anthonsen MW. Toll-like receptor 3 associates with c-Src tyrosine kinase on endosomes to initiate antiviral signaling. 2006. The EMBO Journal. 2006;**25**(14):3335-3346

[54] Shayakhmetov DM, Di Paolo NC, Mossman KL. Recognition of virus infection and innate host responses to viral gene therapy vectors. Molecular Therapy. 2010;**18**(8):1422-1429

[55] Crozat K, Beutler B. TLR7: A new sensor of viral infection. Proceedings of the National Academy of Sciences of the United States of America. 2004;**101**(18):6835-6836

[56] Lam E, Stein S, Falck-Pedersen E. Adenovirus detection by the cGAS/STING/TBK1 DNA sensing cascade. Journal of Virology. 2014;**88**(2):974-981

[57] Orazio N, Naeger C, Karlseder J. Weitzman. The adenovirus E1b55K/E4orf6 complex induces degradation of the Bloom helicase during infection. Journal of Virology. 2010;**85**(4):1887-1892

[58] Misawa T, Takahama M, Kozaki T, Lee H, Zou J, Saitoh T, et al. Microtubule-driven spatial arrangement of mitochondria promotes activation of the NLRP3 inflammasome. Nature Immunology. 2013;**14**:454-460

[59] Fink S, Cookson B. Apoptosis, pyroptosis, and necrosis: Mechanistic description of dead and dying eukaryotic cells. Infection and Immunity. 2005;**73**(4):1907-1916

[60] Berk A. Functions of adenovirus E1A. Cancer Surveys. 1986;**5**(2):367-387

[61] Berk A. Adenovirus promoters and E1A transactivation. Annual Review of Genetics. 1986;**20**:45-79

[62] Kalvakolanu D, Bandyopadhyay S, Harter M, Sen G. Inhibition of interferon-inducible gene expression by adenovirus E1A proteins: Block in transcriptional complex formation. Proceedings of the National Academy of Sciences. 1991;**88**(17):7459-7463

[63] Ackrill A, Foster G, Laxton C, Flavell D, Stark G, Kerr I. Inhibition of the cellular response to interferons by products of the adenovirus type 5 E1A oncogene. Nucleic Acids Research. 1991;**19**(16):4387-4393

[64] Gutch M, Reich N. Repression of the interferon signal transduction pathway by the adenovirus E1A oncogene. Proceedings of the National Academy of Sciences of the United States of America. 1991;**88**(18):7913-7917

[65] Fonseca G, Cohen M, Nichols A, Barrett J, Mymryk J. Viral retasking of hBre1/RNF20 to recruit hPaf1 for transcriptional activation. PLoS Pathogens. 2013;**9**(6):e1003411

[66] Fonseca G, Thillainadesan G, Yousef A, Ablack J, Mossman K, Torchia J, et al. Adenovirus evasion of interferon-mediated innate immunity by direct antagonism of a cellular histone posttranslational modification. Cell Host & Microbe. 2012;**11**(6):597-606

[67] Ferreon AM, Ferreon JC, Wright PE, Deniz AA. Modulation of allostery by protein intrinsic disorder. Nature. 2013;**498**(7454):390

[68] Teodoro J, Branton P. Regulation of p53-dependent apoptosis, transcriptional repression, and cell transformation by phosphorylation of the 55-kilodalton E1B protein of human adenovirus type 5. Journal of Virology. 1997;**71**(5):3620-3627

[69] Sabbatini P, Chiou S, Rao L, White E. Modulation of p53-mediated transcriptional repression and apoptosis by the adenovirus E1B 19K protein. Molecular and Cellular Biology. 1995;**15**(2):1060-1070

[70] Chahal J, Flint S. Timely synthesis of the adenovirus type 5 E1B 55-kilodalton protein is required for efficient genome replication in normal human cells. Journal of Virology. 2012;**86**(6):3064-3072

[71] Chahal J, Qi J, Flint S. The human adenovirus type 5 E1B 55 kDa protein obstructs inhibition of viral replication by type I interferon in normal human cells. PLoS Pathogens. 2012;**8**(8):e1002853

[72] Schreiner S, Bürck C, Glass M, Groitl P, Wimmer P, Kinkley S, et al. Control of human adenovirus type 5 gene expression by cellular Daxx/ATRX chromatin-associated complexes. Nucleic Acids Research. 2013;**41**(6):3532-3550

[73] Schreiner S, Wimmer P, Sirma H, Everett R, Blanchette P, Groitl P, et al. Proteasome-dependent degradation of Daxx by the viral E1B-55K protein in human adenovirus-infected cells. Journal of Virology. 2010;**84**(14):7029-7038

[74] Stracker T, Lee D, Carson C, Araujo F, Ornelles D, Weitzman M. Serotype-specific reorganization of the Mre11 complex by adenoviral E4orf3 proteins. Journal of Virology. 2005;**79**(11):6664-6673

[75] Stracker T, Carson C. Weitzman. Adenovirus oncoproteins inactivate the Mre11–Rad50–NBS1 DNA repair complex. Nature. 2002;**418**:348-352

[76] Evans J, Hearing P. Relocalization of the Mre11-Rad50-Nbs1 complex by the adenovirus E4 ORF3 protein is required for viral replication. Journal of Virology. 2005;**79**(10): 6207-6215

[77] Evans J, Hearing P. Distinct roles of the adenovirus E4 ORF3 protein in viral DNA replication and inhibition of genome concatenation. Journal of Virology. 2003;**77**(9):5295-5304

[78] McSharry B, Burgert H, Owen D, Stanton R, Prod'homme V, Sester M, et al. Adenovirus E3/19K promotes evasion of NK cell recognition by intracellular sequestration of the NKG2D ligands major histocompatibility complex class I chain-related proteins A and B. Journal of Virology. 2008;**82**(9):4585-4594

[79] Cianciola N, Carlin C. Adenovirus RID-alpha activates an autonomous cholesterol regulatory mechanism that rescues defects linked to Niemann-Pick disease type C. The Journal of Cell Biology. 2009;**187**(4):537-552

[80] Fausther-Bovendo H, Kobinger G. Pre-existing immunity against Ad vectors: Humoral, cellular, and innate response, what's important? Human Vaccines & Immunotherapeutics. 2014;**10**(10):2875-2884

[81] Zhang S, Huang W, Zhou X, Zhao Q, Wang Q, Jia B. Seroprevalence of neutralizing antibodies to human adenoviruses type-5 and type-26 and chimpanzee adenovirus type-68 in healthy Chinese adults. Journal of Medical Virology. 2013;**85**(6):1077-1084

[82] Sumida, Truitt D, Lemckert A, Vogels R, JHH C, Addo M, et al. Neutralizing antibodies to adenovirus serotype 5 vaccine vectors are directed primarily against the adenovirus hexon protein. Journal of Immunology (Baltimore, Md: 1950). 2005;**174**(11):7179-7185.

[83] Roberts D, Nanda A, Havenga M, Abbink P, Lynch D, Ewald BA, et al. Hexon-chimaeric adenovirus serotype 5 vectors circumvent pre-existing anti-vector immunity. Nature. 2006;**441**(7090):239-243

[84] Olive M, Eisenlohr L, Flomenberg N, Hsu S, Flomenberg P. The adenovirus capsid protein hexon contains a highly conserved human CD4+ T-cell epitope. Human Gene Therapy. 2002;**13**(10):1167-1178

[85] Roy S, Shirley P, McClelland A, Kaleko M. Circumvention of immunity to the adenovirus major coat protein hexon. Journal of Virology. 1998;**72**(8):6875-6879

[86] Roy S, Clawson D, Calcedo R, Lebherz C, Sanmiguel J, Wu D, et al. Use of chimeric adenoviral vectors to assess capsid neutralization determinants. Virology. 2005;**333**(2):207-214

[87] Calcedo R, Vandenberghe L, Roy S, Somanathan S, Wang L, Wilson J. Host immune responses to chronic adenovirus infections in human and nonhuman primates. Journal of Virology. 2009;**83**(6):2623-2631

[88] Sumida SM, Truitt D, Kishko M, Arthur J, Jackson S, Gorgone D, et al. Neutralizing antibodies and CD8+ T lymphocytes both contribute to immunity to adenovirus serotype 5 vaccine vectors. Journal of Virology. 2004;**78**(6):2666-2673

[89] Veltrop-Duits L, Heemskerk B, Sombroek C, Vreeswijk T, Gubbels S, REE T, et al. Human CD4+ T cells stimulated by conserved adenovirus 5 hexon peptides recognize cells infected with different species of human adenovirus. European Journal of Immunology. 2006;**36**(9):2410-2423

[90] Leen A, Sili U, Vanin E, Jewell A, Xie W, Vignali D, et al. Conserved CTL epitopes on the adenovirus hexon protein expand subgroup cross-reactive and subgroup-specific CD8+ T cells. Blood. 2004;**104**(8):2432-2440

[91] Singh S, Vedi S, Samrat S, Li W, Kumar R, Agrawal B. Heterologous immunity between adenoviruses and hepatitis C virus: A new paradigm in HCV immunity and vaccines. PLoS One. 2016;**11**(1):e0146404

[92] McGrory W, Bautista D, Graham F. A simple technique for the rescue of early region I mutations into infectious human adenovirus type 5. Virology. 1988;**163**(2):614-617

[93] Akusjärvi G. Proteins with transcription regulatory properties encoded by human adenoviruses. Trends in Microbiology. 1993;**1**(5):163-170

[94] Graham FL, Smiley J, Russell WC, Nairn R. Characteristics of a human cell line transformed by DNA from human adenovirus type 5. The Journal of General Virology. 1977;**36**(1):59-74

[95] Shaw G, Morse S, Ararat M, Graham FL. Preferential transformation of human neuronal cells by human adenoviruses and the origin of HEK 293 cells. The FASEB Journal. 2002;**16**(8):869-871

[96] Benihoud K, Yeh P, Perricaudet M. Adenovirus vectors for gene delivery. Current Opinion in Biotechnology. 1999;**10**(5):440-447

[97] Zhang P-X, Fuleihan R. Transfer of activation-dependent gene expression into T cell lines by recombinant adeno-associated virus. Gene Therapy. 1999;**6**(2):3300803

[98] Hehir K, Armentano D, Cardoza L, Choquette T, Berthelette P, White G, et al. Molecular characterization of replication-competent variants of adenovirus vectors and genome modifications to prevent their occurrence. Journal of Virology. 1996;**70**(12):8459-8467

[99] Kovesdi I, Hedley SJ. Adenoviral producer cells. Viruses. 2010;**2**(8):1681-1703

[100] Fallaux FJ, Bout A, van der Velde I, van den Wollenberg DJ, Hehir KM, Keegan J, et al. New helper cells and matched early region 1-deleted adenovirus vectors prevent generation of replication-competent adenoviruses. Human Gene Therapy. 1998;**9**(13):1909-1917

[101] Wold W, Tollefson A, Hermiston T. E3 transcription unit of adenovirus. Current Topics in Microbiology and Immunology. 1995;**199**(1):237-274

[102] Wold W, Doronin K, Toth K, Kuppuswamy M, Lichtenstein DL, Tollefson AE. Immune responses to adenoviruses: Viral evasion mechanisms and their implications for the clinic. Current Opinion in Immunology. 1999;**11**(4):380-386

[103] Bett A, Haddara W, Prevec L, Graham F. An efficient and flexible system for construction of adenovirus vectors with insertions or deletions in early regions 1 and 3. Proceedings of the National Academy of Sciences of the United States of America. 1994;**91**(19):8802-8806

[104] Yang Y, Nunes F, Berencsi K, Gönczöl E, Engelhardt J, Wilson J. Inactivation of E2a in recombinant adenoviruses improves the prospect for gene therapy in cystic fibrosis. Nature Genetics. 1994;**7**:362-369

[105] Yang Y, Nunes F, Berencsi K, Furth E, Gönczöl E, Wilson J. Cellular immunity to viral antigens limits E1-deleted adenoviruses for gene therapy. Proceedings of the National Academy of Sciences of the United States of America. 1994;**91**(10):4407-4411

[106] Engelhardt JF, Litzky L, Wilson JM. Prolonged transgene expression in cotton rat lung with recombinant adenoviruses defective in E2a. Human Gene Therapy. 1994;**5**(10): 1217-1229

[107] Lusky M, Christ M, Rittner K, Dieterle A, Dreyer D, Mourot B, et al. In vitro and in vivo biology of recombinant adenovirus vectors with E1, E1/E2A, or E1/E4 deleted. Journal of Virology. 1998;**71**(3):2022-2032

[108] Schaack J. Adenovirus vectors deleted for genes essential for viral DNA replication. Frontiers in Bioscience: A Journal and Virtual Library. 2005;**1**(10):1146-1155

[109] Bett A, Krougliak V, Graham F. DNA sequence of the deletion/insertion in early region 3 of Ad5 dl309. Virus Research. 1995;**30**(1):75-82

[110] Wang Q, Finer M. Second-generation adenovirus vectors. Nature Medicine. 1996;**2**: 714-716

[111] Kochanek S, Clemens P, Mitani K, Chen H, Chan S, Caskey C. A new adenoviral vector: Replacement of all viral coding sequences with 28 kb of DNA independently expressing both full-length dystrophin and beta-galactosidase. Proceedings of the National Academy of Sciences. 1996;**93**(12):5731-5736

[112] Ishizaki M, Kawano R, Uchida Y, Kimura E, Uchino M, Maeda Y. Gene therapy for duchenne muscular dystrophy by the helper-dependent adenovirus vector (HDAdv)—Mediated full-length dystrophin expression. Molecular Therapy. 2018;**15** (Supplement 1):S52

[113] Puntel M, Muhammad A, Candolfi M, Salem A, Yagiz K, Farrokhi C, et al. A novel bicistronic high-capacity gutless adenovirus vector that drives constitutive expression of herpes simplex virus type 1 thymidine kinase and tet-inducible expression of Flt3L for glioma therapeutics. Journal of Virology. 2010;**84**(12):6007-6017

[114] Parks R, Chen L, Anton M, Sankar U, Rudnicki M, Graham F. A helper-dependent adenovirus vector system: Removal of helper virus by Cre-mediated excision of the viral packaging signal. Proceedings of the National Academy of Sciences of the United States of America. 1996;**93**(24):13565-13570

[115] Fisher KJ, Choi H, Burda J, Chen S-J, Wilson J. Recombinant adenovirus deleted of all viral genes for gene therapy of cystic fibrosis. Virology. 1996;**217**(1):11-22

[116] Hardy S, Kitamura M, Harris-Stansil T, Dai Y, Phipps M. Construction of adenovirus vectors through Cre-lox recombination. Journal of Virology. 1997;**71**(3):1842-1849

[117] Hartigan-O'Connor D, Amalfitano A, Chamberlain J. Improved production of gutted adenovirus in cells expressing adenovirus preterminal protein and DNA polymerase. Journal of Virology. 1999;**73**(9):7835-7841

[118] Ng P, Beauchamp C, Evelegh C, Parks R, Graham F. Development of a FLP/frt system for generating helper-dependent adenoviral vectors. Molecular Therapy: The Journal of the American Society of Gene Therapy. 2001;**3**(5):809-815

[119] Alba R, Hearing P, Bosch A, Chillon M. Differential amplification of adenovirus vectors by flanking the packaging signal with attB/attP-PhiC31 sequences: Implications for helper-dependent adenovirus production. Virology. 2007;**367**(1):51-58

[120] Schiedner G, Morral N, Parks RJ, Wu Y, Koopmans SC, Langston C, et al. Genomic DNA transfer with a high-capacity adenovirus vector results in improved in vivo gene expression and decreased toxicity. Nature Genetics. 1998;**18**(2):180-183

[121] Morral N, Parks R, Zhou H, Langston C, Schiedner G, Quinones J, et al. High doses of a helper-dependent adenoviral vector yield supraphysiological levels of alpha1-antitrypsin with negligible toxicity. Human Gene Therapy. 1998;**9**(18):2709-2716

[122] Chen H, Mack L, Kelly R, Ontell M, Kochanek S, Clemens P. Persistence in muscle of an adenoviral vector that lacks all viral genes. Proceedings of the National Academy of Sciences of the United States of America. 1997;**94**(5):1645-1650

[123] Palmer D, Ng P. Improved system for helper-dependent adenoviral vector production. Molecular Therapy: The Journal of the American Society of Gene Therapy. 2003;**8**(5): 846-852

[124] Lemarchand P, Jaffe H, Danel C, Cid M, Kleinman H, Stratford-Perricaudet L, et al. Adenovirus-mediated transfer of a recombinant human alpha 1-antitrypsin cDNA to human endothelial cells. Proceedings of the National Academy of Sciences. 1992;**89**(14):6482-6486

[125] Davies JC, Geddes DM, Alton E. Prospects for gene therapy in lung disease. Current Opinion in Pharmacology. 2001;**1**(3):272-278

[126] Griesenbach U, Ferrari S, Geddes D, Alton E. Gene therapy progress and prospects: Cystic fibrosis. Gene Therapy. 2002;**9**(20):3301791

[127] Raper S, Wilson J, Yudkoff M, Robinson M, Ye X, Batshaw M. Developing adenoviral-mediated in vivo gene therapy for ornithine transcarbamylase deficiency. Journal of Inherited Metabolic Disease. 1998;**21**(Suppl 1):119-137

[128] Batshaw M, Wilson J, Raper S, Yudkoff M, Robinson M. Recombinant adenovirus gene transfer in adults with partial ornithine transcarbamylase deficiency (OTCD). Human Gene Therapy. 1999;**10**(14):2419-2437

[129] Bramson J, Hitt M, Gauldie J, Graham F. Pre-existing immunity to adenovirus does not prevent tumor regression following intratumoral administration of a vector expressing IL-12 but inhibits virus dissemination. Gene Therapy. 1997;**4**(10):3300508

[130] Sullivan NJ, Sanchez A, Rollin PE, Yang Z, Nabel GJ. Development of a preventive vaccine for Ebola virus infection in primates. Nature. 2000;**408**(6812):605-609

[131] Corey L, McElrath JM, Kublin JG. Post-Step modifications for research on HIV vaccines. AIDS. 2009;**23**(1):3

[132] Shiver JW, Fu T-M, Chen L, Casimiro DR, Davies M-E, Evans RK, et al. Replication-incompetent adenoviral vaccine vector elicits effective anti-immunodeficiency-virus immunity. Nature. 2002;**415**(6869):415331a

[133] Catanzaro AT, Koup RA, Roederer M, Bailer RT, Enama ME, Moodie Z, et al. Phase 1 safety and immunogenicity evaluation of a multiclade HIV-1 candidate vaccine delivered by a replication-defective recombinant adenovirus vector. The Journal of Infectious Diseases. 2006;**194**(12):1638-1649

[134] Gómez-Román RV, Florese RH, Peng B, Montefiori DC, Kalyanaraman VS, Venzon D, et al. An adenovirus-based HIV subtype B prime/boost vaccine regimen elicits antibodies mediating broad antibody-dependent cellular cytotoxicity against non-subtype B HIV strains. JAIDS Journal of Acquired Immune Deficiency Syndromes. 2006;**43**(3):270

[135] Fitzgerald D, Janes H, Robertson M, Coombs R, Frank I, Gilbert P, et al. An Ad5-vectored HIV-1 vaccine elicits cell-mediated immunity but does not affect disease progression in HIV-1–infected male subjects: Results from a randomized placebo-controlled trial (The Step Study). The Journal of Infectious Diseases. 2011;**203**(6):765-772

[136] O'Brien KL, Liu J, King SL, Sun Y-H, Schmitz JE, Lifton MA, et al. Adenovirus-specific immunity after immunization with an Ad5 HIV-1 vaccine candidate in humans. Nature Medicine. 2009;**15**(8):873-875

[137] Harro C, Sun X, Stek JE, Leavitt RY, Mehrotra DV, Wang F, et al. Safety and immunogenicity of the merck adenovirus serotype 5 (MRKAd5) and MRKAd6 human immunodeficiency virus type 1 trigene vaccines alone and in combination in healthy adults. Clinical and Vaccine Immunology. 2009;**16**(9):1285-1292

[138] Patterson L. The "STEP-wise" future of adenovirus-based HIV vaccines. Current Medicinal Chemistry. 2011;**18**(26):3981-3986

[139] Richie TL, Villasante EF. Use of adenovirus serotype 5 vaccine vectors in seropositive, uncircumcised men: Safety lessons from the step trial. The Journal of Infectious Diseases. 2013;**207**(4):689-690

[140] Gray G, Buchbinder S, Duerr A. Overview of STEP and Phambili trial results: Two phase IIb test-of-concept studies investigating the efficacy of MRK adenovirus type 5 gag/pol/nef subtype B HIV vaccine. Current Opinion in HIV and AIDS. 2010;**5**(5):357-361

[141] Michael NL, Robb ML. Phambili: Moving forward without the blindfold. The Lancet Infectious Diseases. 2014;**14**(5):361-362

[142] Pine SO, Kublin JG, Hammer SM, Borgerding J, Huang Y, Casimiro DR, et al. Pre-existing adenovirus immunity modifies a complex mixed Th1 and Th2 cytokine response to an Ad5/HIV-1 vaccine candidate in humans. PLoS One. 2011;**6**(4):e18526

[143] Churchyard GJ, Morgan C, Adams E, Hural J, Graham BS, Moodie Z, et al. A phase IIA randomized clinical trial of a multiclade HIV-1 DNA prime followed by a multiclade rAd5 HIV-1 vaccine boost in healthy adults (HVTN204). PLoS One. 2011;**6**(8):e21225

[144] Janes HE, Cohen KW, Frahm N, Rosa SC, Sanchez B, Hural J, et al. Higher T-cell responses induced by DNA/rAd5 HIV-1 preventive vaccine are associated with lower HIV-1 infection risk in an efficacy trial. The Journal of Infectious Diseases. 2017; **215**(9):1376-1385

[145] Omosa-Manyonyi G, Mpendo J, Ruzagira E, Kilembe W, Chomba E, Roman F, et al. A phase I double blind, placebo-controlled, randomized study of the safety and immunogenicity of an adjuvanted HIV-1 Gag-Pol-Nef fusion protein and adenovirus 35 Gag-RT-Int-Nef vaccine in healthy HIV-uninfected African adults. PLoS One. 2015; **10**(5):e0125954

[146] Keefer MC, Gilmour J, Hayes P, Gill D, Kopycinski J, Cheeseman H, et al. A phase I double blind, placebo-controlled, randomized study of a multigenic HIV-1 adenovirus subtype 35 vector vaccine in healthy uninfected adults. PLoS One. 2012;**7**(8):e41936

[147] Baden LR, Karita E, Mutua G, Bekker L-G, Gray G, Page-Shipp L, et al. Assessment of the safety and immunogenicity of 2 novel vaccine platforms for HIV-1 prevention: A randomized trial. Annals of Internal Medicine. 2016;**164**(5):313-322

[148] Baden LR, Walsh SR, Seaman MS, Tucker RP, Krause KH, Patel A, et al. First-in-human evaluation of the safety and immunogenicity of a recombinant adenovirus serotype 26 HIV-1 Env vaccine (IPCAVD 001). The Journal of Infectious Diseases. 2013;**207**(2):240-247

[149] Baden LR, Walsh SR, Seaman MS, Cohen YZ, Johnson JA, Licona HJ, et al. First-in-human randomized controlled trial of mosaic HIV-1 immunogens delivered via a modified vaccinia Ankara vector. The Journal of Infectious Diseases. 2018;**218**(4):633-644

[150] Paris R, Kuschner RA, Binn L, Thomas SJ, Colloca S, Nicosia A, et al. Adenovirus Type 4 and 7 vaccination or adenovirus Type 4 respiratory infection elicits minimal cross-reactive antibody responses to nonhuman adenovirus vaccine vectors. Clinical and Vaccine Immunology. 2014;**21**(5):783-786

[151] Moffatt S, Hays J, HogenEsch H, Mittal S. Circumvention of vector-specific neutralizing antibody response by alternating use of human and non-human adenoviruses: Implications in gene therapy. Virology. 2000;**272**(1):159-167

[152] Earl PL, Americo JL, Wyatt LS, Eller L, Montefiori DC, Byrum R, et al. Recombinant modified vaccinia virus Ankara provides durable protection against disease caused by an immunodeficiency virus as well as long-term immunity to an orthopoxvirus in a non-human primate. Virology. 2007;**366**(1):84-97

[153] Tapia MD, Sow SO, Lyke KE, Haidara F, Diallo F, Doumbia M, et al. Use of ChAd3-EBO-Z Ebola virus vaccine in Malian and US adults, and boosting of Malian adults with MVA-BN-Filo: A phase 1, single-blind, randomised trial, a phase 1b, open-label and

double-blind, dose-escalation trial, and a nested, randomised, double-blind, placebo-controlled trial. The Lancet Infectious Diseases. 2016;**16**(1):31-42

[154] Milligan ID, Gibani MM, Sewell R, Clutterbuck EA, Campbell D, Plested E, et al. Safety and immunogenicity of novel adenovirus type 26- and modified vaccinia Ankara–vectored Ebola vaccines: A randomized clinical trial. Journal of the American Medical Association. 2016;**315**(15):1610-1623

[155] Wu S, Kroeker A, Wong G, He S, Hou L, Audet J, et al. An adenovirus vaccine expressing Ebola virus variant makona glycoprotein is efficacious in Guinea pigs and nonhuman primates. The Journal of Infectious Diseases. 2016;**214**(suppl_3):S326-S332

[156] Kennedy SB, Bolay F, Kieh M, Grandits G, Badio M, Ballou R, et al. Phase 2 placebo-controlled trial of two vaccines to prevent Ebola in Liberia. The New England Journal of Medicine. 2017;**377**(15):1438-1447

[157] Santis O, Audran R, Pothin E, Warpelin-Decrausaz L, Vallotton L, Wuerzner G, et al. Safety and immunogenicity of a chimpanzee adenovirus-vectored Ebola vaccine in healthy adults: A randomised, double-blind, placebo-controlled, dose-finding, phase 1/2a study. The Lancet Infectious Diseases. 2016;**16**(3):311-320

[158] Wu L, Zhang Z, Gao H, Li Y, Hou L, Yao H, et al. Open-label phase I clinical trial of Ad5-EBOV in Africans in China. Human Vaccines & Immunotherapeutics. 2017;**13**(9):2078-2085

[159] Ogwang C, Afolabi M, Kimani D, Jagne Y, Sheehy SH, Bliss CM, et al. Safety and immunogenicity of heterologous prime-boost immunisation with *Plasmodium falciparum* malaria candidate vaccines, ChAd63 ME-TRAP and MVA ME-TRAP, in healthy Gambian and Kenyan Adults. PLoS One. 2013;**8**(3):e57726

[160] Jhawar SR, Thandoni A, Bommareddy PK, Hassan S, Kohlhapp FJ, Goyal S, et al. Oncolytic viruses—Natural and genetically engineered cancer immunotherapies. Frontiers in Oncology. 2017;**7**:202

[161] Warner SG, O'Leary MP, Fong Y. Therapeutic oncolytic viruses: Clinical advances and future directions. Current Opinion in Oncology. 2017;**29**(5):359

[162] Shaw A, Suzuki M. Recent advances in oncolytic adenovirus therapies for cancer. Current Opinion in Virology. 2016;**21**:9-15

[163] Wold W, Toth K. Adenovirus vectors for gene therapy, vaccination and cancer gene therapy. Current Gene Therapy. 2014;**13**(6):421-433

[164] Yu W, Fang H. Clinical trials with oncolytic adenovirus in China. Current Cancer Drug Targets. 2007;**7**(2):141-148

[165] Cerullo V, Capasso C, Vähä-Koskela M, Hemminki O, Hemminki A. Cancer-targeted oncolytic adenoviruses for modulation of the immune system. Current Cancer Drug Targets. 2018;**18**(2):124-138

[166] Yamamoto Y, Nagasato M, Yoshida T, Aoki K. Recent advances in genetic modification of adenovirus vectors for cancer treatment. Cancer Science. 2017;**108**(5):831-837

[167] Cerullo V, Vähä-Koskela M, Hemminki A. Oncolytic adenoviruses. OncoImmunology. 2012;**1**(6):979-981

[168] Cerullo V, Diaconu I, Romano V, Hirvinen M, Ugolini M, Escutenaire S, et al. An oncolytic adenovirus enhanced for toll-like receptor 9 stimulation increases antitumor immune responses and tumor clearance. Molecular Therapy. 2012;**20**(11):2076-2086

[169] Heise C, Williams A, Xue S, Propst M, Kirn D. Intravenous administration of ONYX-015, a selectively replicating adenovirus, induces antitumoral efficacy. Cancer Research. 1999;**59**(11):2623-2628

[170] Khuri FR, Nemunaitis J, Ganly I, Arseneau J, Tannock IF, Romel L, et al. A controlled trial of intratumoral ONYX-015, a selectively-replicating adenovirus, in combination with cisplatin and 5-fluorouracil in patients with recurrent head and neck cancer. Nature Medicine. 2000;**6**(8):nm0800_879

[171] Räty J, Pikkarainen J, Wirth T, Ylä-Herttuala S. Gene therapy: The first approved gene-based medicines, molecular mechanisms, and clinical indications. Current Molecular Pharmacology. 2008;**1**(1):13-23

[172] Guo W, Song H. Development of gene therapeutics for head and neck cancer in China: From bench to bedside. Human Gene Therapy. 2018;**29**(2):180-187

[173] Sangro B, Mazzolini G, Ruiz M, Ruiz J, Quiroga J, Herrero I, et al. A phase I clinical trial of thymidine kinase-based gene therapy in advanced hepatocellular carcinoma. Cancer Gene Therapy. 2010;**17**(12):837

[174] Zhan Y, Yu B, Wang Z, Zhang Y, Zhang H-H, Wu H, et al. A fiber-modified adenovirus co-expressing HSV-TK and Coli.NTR enhances antitumor activities in breast cancer cells. International Journal of Clinical and Experimental Pathology. 2014;**7**(6):2850-2860

[175] Barton KN, Paielli D, Zhang Y, Koul S, Brown SL, Lu M, et al. Second-generation replication-competent oncolytic adenovirus armed with improved suicide genes and ADP gene demonstrates greater efficacy without increased toxicity. Molecular Therapy. 2006;**13**(2):347-356

[176] Saito H, Kitagawa K, Yoneda T, Fukui Y, Fujsawa M, Bautista D, et al. Combination of p53-DC vaccine and rAd-p53 gene therapy induced CTLs cytotoxic against p53-deleted human prostate cancer cells in vitro. Cancer Gene Therapy. 2017;**24**(7):289-296

[177] Zhou J, Huang Z, Wang Z, Liu S, Grandien A, Ernberg I, et al. Tumor suppressor BLU promotes TRAIL-induced apoptosis by downregulating NF-κB signaling in nasopharyngeal carcinoma. Oncotarget. 2014;**8**(27):43853-43865

[178] Gabrilovich DI. INGN 201 (Advexin): Adenoviral p53 gene therapy for cancer. Expert Opinion on Biological Therapy. 2006;**6**(8):823-832

[179] Senzer N, Nemunaitis J. A review of contusugene ladenovec (Advexin) p53 therapy. Current Opinion in Molecular Therapeutics. 2009;**11**(1):54-61

[180] Chen G, Zhang S, He X, Liu S, Ma C, Zou X-P. Clinical utility of recombinant adenoviral human p53 gene therapy: Current perspectives. OncoTargets and Therapy. 2014;**7**:1901-1909

[181] Xia Y, Du Z, Wang X, Li X. Treatment of uterine sarcoma with rAd-p53 (Gendicine®) followed by chemotherapy—Clinical study on TP53 gene therapy. Human Gene Therapy. 2018;**29**(2):242-250

[182] Vitelli A, Folgori A, Scarselli E, Colloca S, Capone S, Nicosia A. Chimpanzee adenoviral vectors as vaccines—Challenges to move the technology into the fast lane. Expert Review of Vaccines. 2017;**16**(12):1241-1252

[183] Short J, Pereboev A, Kawakami Y, Virology V. Adenovirus serotype 3 utilizes CD80 (B7. 1) and CD86 (B7. 2) as cellular attachment receptors. Virology. 2004;**322**(2):349-359

[184] Zhang C, Yang Y, Chi Y, Yin J, Yan L, Ku Z, et al. Hexon-modified recombinant E1-deleted adenoviral vectors as bivalent vaccine carriers for Coxsackievirus A16 and Enterovirus 71. Vaccine. 2015;**33**(39):5087-5094

[185] Tuve S, Wang H, Jacobs JD, Yumul RC, Smith DF, Lieber A. Role of cellular heparan sulfate proteoglycans in infection of human adenovirus serotype 3 and 35. PLoS Pathogens. 2008;**4**(10):e1000189

[186] Chen H, Xiang Z, Li Y, Kurupati R, Jia B, Bian A, et al. Adenovirus-based vaccines: Comparison of vectors from three species of adenoviridae. Journal of Virology. 2010;**84**(20):10522-10532

[187] Geisbert TW, Bailey M, Hensley L, Asiedu C, Geisbert J, Stanley D, et al. Recombinant adenovirus serotype 26 (Ad26) and Ad35 vaccine vectors bypass immunity to Ad5 and protect nonhuman primates against Ebolavirus challenge. Journal of Virology. 2011;**85**(9):4222-4233

[188] Barouch DH, Pau MG, Custers JH, Koudstaal W, Kostense S, Havenga MJ, et al. Immunogenicity of recombinant adenovirus serotype 35 vaccine in the presence of pre-existing anti-Ad5 immunity. The Journal of Immunology. 2004;**172**(10):6290-6297

[189] Holterman L, Vogels R, van der Vlugt R, Sieuwerts M, Grimbergen J, Kaspers J, et al. Novel replication-incompetent vector derived from adenovirus type 11 (Ad11) for vaccination and gene therapy: Low seroprevalence and non-cross-reactivity with Ad5. Journal of Virology. 2004;**78**(23):13207-13215

[190] Capone S, Meola A, Ercole B, Vitelli A, Pezzanera M, Ruggeri L, et al. A novel adenovirus type 6 (Ad6)-based hepatitis C virus vector that overcomes preexisting anti-Ad5 immunity and induces potent and broad cellular immune responses in rhesus macaques. Journal of Virology. 2006;**80**(4):1688-1699

[191] Swadling L, Capone S, Antrobus RD, Brown A, Richardson R, Newell EW, et al. A human vaccine strategy based on chimpanzee adenoviral and MVA vectors that primes, boosts, and sustains functional HCV-specific T cell memory. Science Translational Medicine. 2014;**6**(261):261ra153

[192] Biswas S, Choudhary P, Elias SC, Miura K, Milne KH, de Cassan SC, et al. Assessment of humoral immune responses to blood-stage malaria antigens following ChAd63-MVA immunization, controlled human malaria infection and natural exposure. PLoS One. 2014;**9**(9):e107903

[193] Reddy PS, Idamakanti N, Zakhartchouk AN, Baxi MK, Lee JB, Pyne C, Babiuk LA, Tikoo SK. Nucleotide sequence, genome organization, and transcription map of bovine adenovirus type 3. Journal of Virology. 1998;**72**(2):1394-1402

[194] Hammond JM1, McCoy RJ, Jansen ES, Morrissy CJ, Hodgson AL, Johnson MA. Vaccination with a single dose of a recombinant porcine adenovirus expressing the classical swine fever virus gp55 (E2) gene protects pigs against classical swine. Vaccine. 2000;**18**(11-12):1040-1050

[195] Wüest T, Both GW, Prince AM, Hofmann C, Löser P. Recombinant ovine atadenovirus induces a strong and sustained T cell response against the hepatitis C virus NS3 antigen in mice. Vaccine. 2004;**22**(21-22):2717-2721

[196] Xiang Z, Gao G, Reyes-Sandoval A, Li Y, Wilson J, Ertl H. Oral vaccination of mice with adenoviral vectors is not impaired by preexisting immunity to the vaccine carrier. Journal of Virology. 2003;**77**(20):10780-10789

[197] Lin SW, Cun AS, Harris-McCoy K, Ertl HC. Intramuscular rather than oral administration of replication-defective adenoviral vaccine vector induces specific CD8+ T cell responses in the gut. Vaccine. 2007;**25**(12):2187-2193

[198] Mackay LK, Stock AT, Ma JZ, Jones CM, Kent SJ, Mueller SN, et al. Long-lived epithelial immunity by tissue-resident memory T (TRM) cells in the absence of persisting local antigen presentation. Proceedings of the National Academy of Sciences of the United States of America. 2012;**109**(18):7037-7042

[199] Richardson JS, Pillet S, Bello AJ, Kobinger GP. Airway delivery of an adenovirus-based Ebola virus vaccine bypasses existing immunity to homologous adenovirus in nonhuman primates. Journal of virology. 2013;**87**(7):3668-3677

[200] Croyle MA1, Patel A, Tran KN, Gray M, Zhang Y, Strong JE, Feldmann H, Kobinger GP. Nasal delivery of an adenovirus-based vaccine bypasses pre-existing immunity to the vaccine carrier and improves the immune response in mice. PLoS One. 2008;**3**(10):e3548

[201] McCoy K, Tatsis N, Korioth-Schmitz B, Lasaro MO, Hensley SE, Lin S-W, et al. Effect of preexisting immunity to adenovirus human serotype 5 antigens on the immune responses of nonhuman primates to vaccine regimens based on human- or chimpanzee-derived adenovirus vectors. Journal of Virology. 2007;**81**(12):6594-6604

[202] Afolabi MO, Tiono AB, Adetifa UJ, Yaro J, Drammeh A, Nébié I, et al. Safety and immunogenicity of ChAd63 and MVA ME-TRAP in West African children and infants. Molecular Therapy. 2016;**24**(8):1470-1477

[203] Mensah VA, Gueye A, Ndiaye M, Edwards NJ, Wright D, Anagnostou NA, et al. Safety, immunogenicity and efficacy of prime-boost vaccination with ChAd63 and MVA encoding ME-TRAP against *Plasmodium falciparum* infection in adults in Senegal. PLoS One. 2016;**11**(12):e0167951

[204] Hayton E-J, Rose A, Ibrahimsa U, Sorbo M, Capone S, Crook A, et al. Safety and tolerability of conserved region vaccines vectored by plasmid DNA, simian adenovirus and modified vaccinia virus Ankara administered to human immunodeficiency virus type 1-uninfected adults in a randomized, single-blind phase I trial. PLoS ONE. 2014;**9**(7):e101591

[205] Kaulfuß M, Wensing I, Windmann S, Hrycak C, Bayer W. Induction of complex immune responses and strong protection against retrovirus challenge by adenovirus-based immunization depends on the order of vaccine delivery. Retrovirology. 2017;**14**(1):8

[206] Schöne D, Hrycak CP, Windmann S, Lapuente D, Dittmer U, Tenbusch M, et al. Immunodominance of adenovirus-derived CD8+ T cell epitopes interferes with the induction of transgene-specific immunity in adenovirus-based immunization. Journal of Virology. 2017;**91**(20):e01184-17

[207] Schirmbeck R, Reimann J, Kochanek S, Kreppel F. The immunogenicity of adenovirus vectors limits the multispecificity of CD8 T-cell responses to vector-encoded transgenic antigens. Molecular Therapy. 2008;**16**(9):1609-1616

[208] Agrawal B, Singh S, Gupta N, Li W, Vedi S, Kumar R. Unsolved puzzles surrounding HCV immunity: Heterologous immunity adds another dimension. International Journal of Molecular Sciences. 2017;**18**(8):1626-1639

[209] Lee CS et al. Adenovirus mediated gene delivery: Potential applications for gene and cell-based therapies in the new era of personalized medicine. Genes and Diseases. 2017;**4**(2):43-63. DOI: 10.10.1016/j.gendis.2017.04.001